P9-CDK-116

JOURNEYS
IN MICROSPACE

JOURNEYS IN MICROSPACE

THE ART OF THE SCANNING ELECTRON MICROSCOPE

DEE BREGER

COLUMBIA UNIVERSITY PRESS
NEW YORK

CONTENTS

TO THE MEMORIES OF DAVE BREGER
AND DORATHY LEWIS BREGER, MY PARENTS.

ACKNOWLEDGMENTS

JOURNEYS IN MICROSPACE germinated during a research session in which I was operating the SEM under the direction of Lamont tectonophysicist Garry Karner. Conversation turned to my dream of publishing a book of micrographs: over the years, I'd noticed that folks outside the cloistered realm of science enjoyed their glimpses of the microworld through my pictures, and I'd long wanted to share my privileged access on a broader scale. Garry was on a relevant committee; three years later, I gave to Ed Lugenbeel, Executive Editor of Columbia University Press, a stack of prints, some text, and my dream. I thank Garry for making my book happen, and Ed for his enthusiasm and support.

The most important contributors to this project are, of course, the scientists who allowed me use of their samples after their work was done, and who gave me their time and knowledge when I approached them for descriptions of the science contained in my pictures. For them this venture was uncharted territory, but, being explorers, they were willing to gamble with me. I have tried to keep the scientific discussions in their own words, and if any factual errors have crept into the captions or background information, they are from my adjustments of the original material. The contributing scientists are noted individually in the Sources & Stories section at the end of the book, but here I extend to each of them special and heartfelt thanks for their participation and trust.

The Polaroid Corporation and Leica Inc. have generously aided this project, and I most appreciate their support of my work. Thanks also to Dick Daniel of Radco, and the Denton Corporation, whose sample preparation equipment helped ensure that only quality samples went under the beam. Jeff Streger of Leica/Cambridge and Marcel C. Broussard of OSC Technologies are two superior service engineers, and their care of the instrument shows in the micrographs. I am grateful to Andy Leonard for computer color-enhancing several of my black and white prints so attractively, Brad

Hyman for printing the brown-tone micrographs despite being baffled by the subject matter, and Ted Baker for photographing the microscope and the microscopist with patient and cheerful humor. From Columbia University Press, I am indebted to Ivon Katz, Manuscript Editor, and Teresa Bonner, Design Manager, who molded my raw efforts into the polished creation of professionals.

Interweaving science and art without compromising the integrity of either would have been difficult without the clarity of vision of Enrico Bonatti, marine geologist, advisor, and friend. His influence shows throughout the book, and I thank him for pulling me out of some blind alleys and orienting me back in the right direction. I am fortunate also to have publishing professionals in my family: my sister, Lois Breger, and brother, Harry Breger, who have given me expert (and emphatically objective) guidance and feedback throughout this project. Other friends and family members have given me much personal support. Some of them donated samples as well, which appear in these pages to their amusement, if not their surprise. Foremost among these special people are Nancy Howard Gray, Kate Davidson, Kathy Morrison, Ken Wade, Betty Jonas, and Gigi Margel.

Finally, I would like to express my lifelong gratitude to Tsune Saito, micropaleontologist and friend, for originally setting me free on the other side of the electronic looking glass.

INTRODUCTION

SCALES, DIMENSIONS, AND PERSPECTIVES

■

COSMOLOGISTS WHO PONDER THE BIGGEST REALM OF ALL, and particle physicists who study the smallest, both tell us that the universe suddenly sprang into featureless emptiness from a point infinitely small. Everything, it seems, was once contained in nothing.

The universe is incomprehensibly big, yet we tie ourselves up into mental knots by trying to imagine what contains it. On the other end of the scale, our minds boggle by trying to imagine the point of disembodied energy that became the universe. From the beginnings of consciousness, people have grappled with these kinds of thoughts. Based on the observations of their own time and place and the means at their disposal, all cultures have attempted to explain the natural world. Revelations lately grounded in sophisticated technology have merely made us more knowledgeable, not necessarily more enlightened.

From tribal reverence of nature to university research into its mysteries, we remain in awe of the variety of forms manifesting existence. We are equally astonished by its enormous range of scale: advanced telescopes are continually expanding our range of vision, revealing ever more detailed pictures of ever larger cosmic objects. Not so long ago we thought a single galaxy was big, but lately we've learned there are curving sheets of supergalactic clusters hundreds of millions of light-years long.

Atoms were once the smallest things known to us, but physicists have found a galaxy of subatomic particles that in many respects eerily echoes the cosmos at large. Tools like supercolliders have revealed other particles even smaller. Quarks live so briefly that, lying barely on the edge of being, they've scarcely been here before they're gone.

The dimensions of quarks and clusters are difficult to comprehend. On the scale of everyday human experience, however, the unaided eye gives us no trouble in conceptualizing things larger than ourselves: whales, forests, cities, the moon. . . . Our everyday comprehension starts to fade perhaps just past the level of the Milky Way.

We can also easily comprehend things much smaller than we are: owls, roses, buttons, grains of sand. But at the level of a sand grain, another world begins to unfold. This world is normally invisible, yet it's as breathtakingly complex as our familiar environment and as rich as the cosmos revealed by telescopes. It's as beautiful as both, when we stop to pay attention.

Most people don't usually get a chance to make close contact with the research technologies that extend everyday experience. Though we may suspect that the use of instruments like powerful microscopes and telescopes can inspire us to envision a universe of imperceptible interconnections, we normally have to passively accept the scientists' dry results and interpretations.

This book makes accessible a small part of that restricted domain. Within these pages, you can catch a glimpse of an extraordinary microworld. Here you enter the level that begins just beyond the reach of familiar light microscopes, where classroom memories are left behind. This is your invitation to cross to the other side of the magnifying glass, and wander, like Alice, through the world of the scanning electron microscope.

THE SCANNING ELECTRON MICROSCOPE

Known fondly and expediently as the SEM (pronounced either "sem" or "S.E.M."), the scanning electron microscope is the only tool we have that can magnify small objects or minute parts of objects in a way that shows how they would look if we could see them without the microscope.

Ordinary optical microscopes use units of light, called photons, to look directly at samples through the magnifying capability of glass lenses. But photons are large as subatomic particles go, which limits the amount of available magnification to only several thousand times actual size. Another problem is that only one level of the sample can be focused at a time. If the object under the lens has any three-dimensional structure, part of it will always be out of focus. Depending on where you focus, this gives you the choice of either blurred outlines or a blurred middle and often results in an image that doesn't look too much like the original specimen.

Just out of reach of visible light, however, an infinite variety of very small things can be seen at higher magnifications. Since they're too small to be revealed by photons, trying to see them with photons would be like trying to play marbles with cannon balls.

Enter electrons. Not only are they small enough to provide much greater resolving power (that is, they are able to magnify much more effectively), they are also abundant and easily manipulated. When we shoot a steady beam of electrons through electromagnetic lenses onto the surface of a small object and collect electrons that result from the impact, we have a new system that simultaneously solves the problems of limited magnification and depth of field: a scanning electron microscope. Adding a viewing screen and a camera completes the picture.

An electron "gun" aimed down the center of a vertical column is the source of the beam. Applying voltage to a filament inside the gun shoots electrons down through a stack of circular electromagnetic lenses that shape, define, and focus the stream into a coherent beam that finally lands as a fine point on the specimen below. The specimen is held under the electron beam in a vacuum chamber at the bottom of the column, on a device with external controls that can tilt, rotate, and move it forward, backward, up, and down.

The electrons in the beam (called primary electrons) strike the specimen in a constant scanning motion and cause it to eject some of its own electrons (called secondary electrons), which carry information about the structure of its surface. These secondary electrons are collected by a specialized detector inside the chamber, electronically processed, and instantaneously reassembled in their original configuration on a cathode ray tube (CRT), a viewing screen similar to a television screen. The

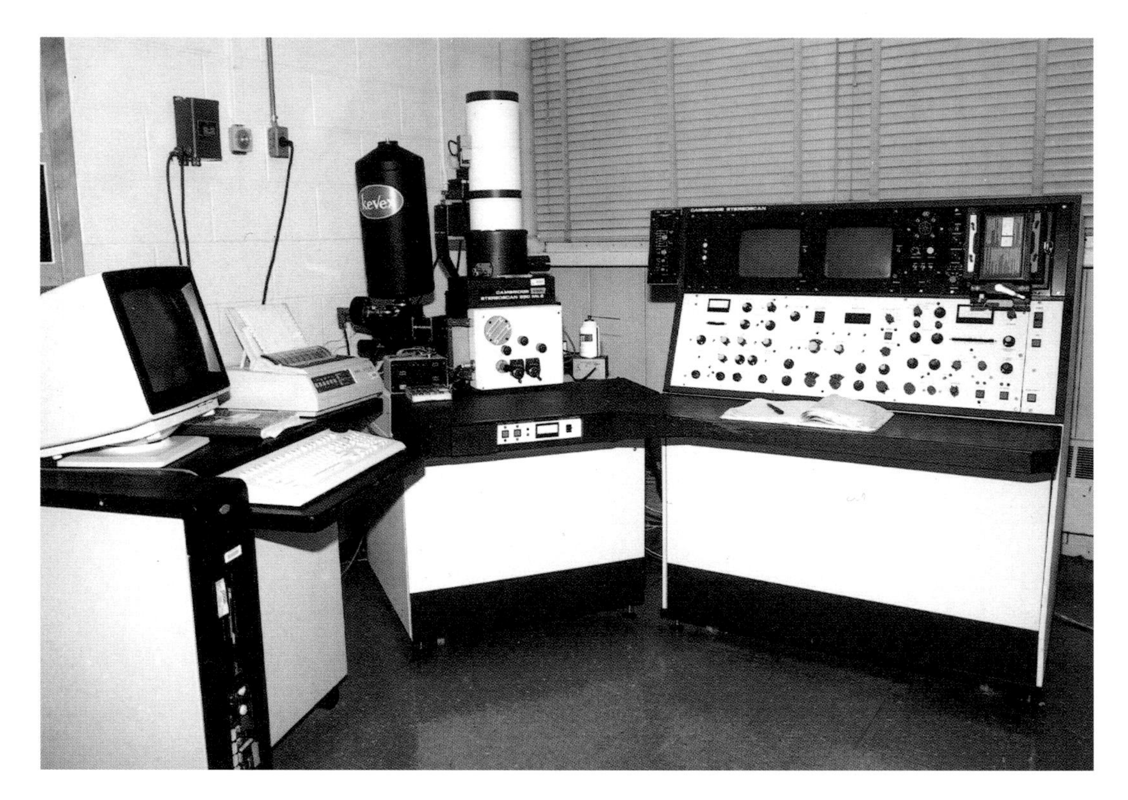

1.1 Lamont's Scanning Electron Microscope

CRT displays a picture of the sample's surface that changes continuously as the operator continues to explore.

On a separate console next to the column is an electronic panel with knobs, dials, and switches that control the beam, the lenses, the various detectors inside the sample chamber, and the image on one or more CRTs. The magnification control causes the picture to zoom in and out, showing first the contours of the whole specimen at perhaps ten or twenty times its actual size, then flying in for a closer look at details thousands or tens of thousands times actual size. The operator can wander over the surface at any "elevation," turn the sample around, move and tilt it as desired, and hover for a longer look or a photo.

By moving the sample and changing magnification, focus, and beam parameters, a sample can be explored and photographed with a clarity of detail and depth of focus unavailable with any other technology. Magnifications well in excess of 300,000 times can be achieved. (Consider: at a magnification of 300,000, your thumbnail would appear as wide as Manhattan Island is in reality. Conversely, if Manhattan were condensed so that the distance between the Hudson and the East River was the width of your thumbnail, the SEM could study the city brick by brick. Most micrographs are taken at magnifications of less than 50,000 times, however, where there is a greater wealth of meaningful structure. In these ranges we could use the SEM to read the street signs, determine which neighborhoods are the most populated, locate restaurants, and leap tall buildings in a single bound.)

When the electron beam encounters the sample, another detector collects some of the primary electrons that had originated in the beam, hit the sample, and bounced back again. Areas on the sample composed of chemical elements of lower atomic number, such as magnesium, silicon, or calcium, will absorb more of the beam's electrons and bounce fewer back toward this detector. Parts containing elements of higher atomic number, such as iron or barium, will absorb fewer primary electrons and scatter back many more. Since each backscattered primary electron contributes a white dot to the image on the CRT, the resulting picture is actually a chemical map, on which brighter and darker parts represent areas of differing chemical composition. A bit of rock will be composed of many minerals, but they can't easily be differentiated in the secondary image; the backscatter image displays them nicely. Due to several factors in the way they're generated, backscatter images appear flatter and not as sharp as secondary images.

There are two ways of preparing a sample for backscatter imaging: by looking at it in its original rough form to reveal the chemical components as they exist naturally, or by polishing a flat slice of the sample to see their planar relationships. With two viewing screens, secondary and backscatter images can be viewed simultaneously. Any image can be recorded with a camera attached to a separate photographic CRT. A few of the pictures in this book are backscatter images, or combinations of secondary and backscatter, in which topographic and chemical information is electronically mixed.

THE MICROGRAPHS

■

Many of the micrographs presented here are from samples used in various scientific research projects in the SEM laboratory of Lamont-Doherty Earth Observatory. Lamont is a world-class Earth Science research institute affiliated with Columbia University, situated on over a hundred acres of the Palisades Cliffs overlooking the Hudson River. About fifteen miles north of the main Columbia campus in New York City, the land originally served as a family estate. In 1948 Florence C. Lamont, the widow of financier Thomas Lamont, donated the grounds to Columbia. The estate was subsequently turned over to the Geology Department, primarily to provide a quiet environment for the development of sensitive seismometers. Since then a great deal of intense scientific research has been conducted in this idyllic setting, and much of what is now known about the Earth has been formulated here.

In the years since the newly available SEM technology came to Lamont in the late 1960s, many on-site researchers have used two successive SEMs to help solve some of the Earth's mysteries in marine and continental geology, micropaleontology, biological oceanography, experimental petrology, paleomagnetics, and sedimentology. Since 1982, researchers from other disciplines in the Columbia community have also traveled to Lamont to examine specimens from fields as diverse as medical research, chemical engineering, materials science, mining, and archaeology. Scientists from both international and other American universities, institutions, and museums have also brought intriguing projects to Lamont's SEM.

Although the SEM is dedicated to experimental investigation, images of arresting intimacy and beauty are often revealed in the course of exploring the subjects of research. Scientists may appreciate the aesthetics of their samples, but they don't have the luxury of stopping to marvel while they work. The approach taken here, however, is aesthetic rather than scientific. Although several of these micrographs were taken during research projects and have been published in scholarly journals or hang in science museums, most of these images were necessarily bypassed by scientists in their timely pursuit of specific answers to specific questions.

On these pages, then, you can see a sampling of what the experts have seen when they could not linger. You might say that these images, lying at the heart of *Journeys in Microspace*, represent the sublime side of science. Other micrographs are views of everyday objects. Certain samples, unexpectedly rich in form and texture, invited more visual exploration than others did, so that in wandering through these pages you will notice that several extended suites of images arose from the same samples. Indeed, it would have been possible to fill a book from nothing more than a few microminerals, a sprinkling of nearly invisible microplankton or a couple of insects. But with so much beauty in the microworld at large, so to speak, these images have been drawn from a wider variety of sources that were always intriguing, sometimes esoteric, occasionally exotic, even mysterious.

Two pictures of the same sample can look completely different, depending on the interplay among various methods of sample preparation, how the microscope was used, photographic settings, film type, and darkroom techniques. Two microscopists using the same instrument and sample will come up with different micrographs, and the same microscopist can vary results in the same ways. Reducing the current (and therefore the diameter) of the electron beam, for instance, induces grain and reduces contrast, as can be seen in Plate 2.18, regardless of the inherent properties of the film being used. Expanding the beam beyond the normal range produces the opposite effect, as in Plates 2.5 and 2.10. Some of the images in the following pages took advantage of procedural accidents somewhere along the line that presented offbeat effects worth exploiting, as in Plate 3.19.

The micrographs in this book are analog images produced with a 1982 Leica/Cambridge Stereoscan 250 Mark 2 scanning electron microscope. This instrument's capabilities and versatility have been proved time and again over the years in its response to the demands of leading-edge research.

Using sensitive SEM technology can be tricky, however. Analog micrographs are slowly time-exposed one scan line after another, and in the time it takes for the camera to record an image it can also record the fact that the sample has reacted to heat from the electron beam or that the focus has drifted; unacceptable artifacts can be created in many ways. Each picture also requires its own photographic settings. Most of these micrographs were taken on Polaroid Type 55 black-and-white instant film (which allows an operator to monitor each image as it's created and gives him or her the possibility for subsequent electronic or photographic fine-tuning).

Electron micrographs are initially black-and-white because the samples' real colors are negated by electron imaging. The blue-tone pictures in this book were taken on Polaroid Presentation Chrome 35 mm color slide film, which simply recorded the blue scan line of the photographic CRT; the color of the micrograph has nothing to do with the color of the sample. The brown-tone pictures were taken on Ilford XP-1 monochromatic 35 mm film. Some black-and-white prints were computer color-enhanced by Andrew Paul Leonard, whose color choices were inspired by the moment.

Of the many groups of microplankton, the four represented in these pages (radiolaria, coccolithophyceae, diatoms, and foraminifera) are described in the Sources & Stories section at the end of the book by some of the Lamont scientists who study them. Captions can only hint at the complex backgrounds of many of the images on the following pages, and these stories are also told there in further detail. The number in parentheses that accompanies each caption refers to the note(s) telling the story of that image.

Finally, it must be stressed that these are all photographs of real objects. In the age of "virtual reality" and other computer-generated imagery, we need to keep in mind that these electron micrographs show us a reality created in the natural world with no particular regard to our presence and captured only by our intrusion. Even the images derived from experiments show materials operating under natural laws. We can only observe, fascinated.

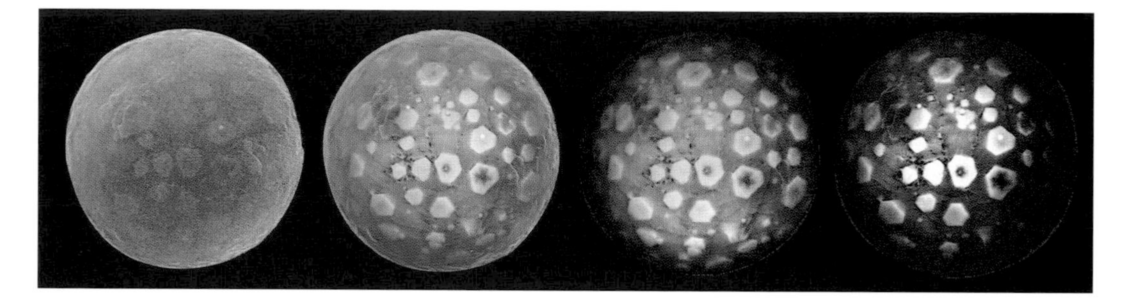

1.2 Microspherule, possibly of cosmic origin, x 1,700 (5)

PITFALLS AND PARADOXES

Scanning electron microscopes don't merely use electron beams to illuminate objects so small they can't be seen by light. Since they are electronic devices, SEMs can manipulate isolated groups of electrons from the sample-beam interaction to create separate pictures (such as secondary and backscatter images) that contain different kinds of information about a single object. This variety can lead to a "compound reality" or, since no version tells the whole story, a kind of ironic nonreality. And yet any image created with an SEM is still real in the sense that computer-generated images are not.

In another sense, the object that looks so solid on the screen or in the micrograph doesn't exist at all. Only electronic cables connect the image on the screen with the object in the sample chamber. In other words, though we experience the effect of direct observation, we are not even looking at the object we see. Electron microscopy is indirect and so can be disorienting. It's a matter of illusions echoing across dimensions, of technology-induced surreality.

Accepting, however, that an SEM displays reality on a practical level and that anything revealed does exist, using the instrument admits us to a rich, otherwise invisible world. Of course, the microworld has always secretly existed parallel to our own, but before the first microscopes began to be developed in the seventeenth century, its intricacy was unimagined. Now that we have extremely powerful methods of magnification, we are continually astounded by the microworld's complexity, mystery, and beauty.

Magnified views of a feather reveal a microfractal sequence. On a given scale, too, forms repeat endlessly from one material, one context, to another. But nature also repeats unrelated shapes from one dimension to another, and forms found on grand scales are echoed in intermediate ones and found again on the microscopic level. Circles and spheres, cubes and squares, cones and triangles, stars, ripples, branches, spirals, and similar shapes and figures are infinitely replicated in the macroworld, and we see them again in the SEM.

Images produced by an SEM, isolated from context and out of proportion, can therefore become oddly evocative of the macroworld. Magnification is expressed only as a number, and we miss visual clues to the scale of these pictures; the resulting disorientation is a little unsettling. With the same compulsion that created the constellations, we find ourselves linking these micrographs to objects more comfortably familiar, however unrelated they may actually be.

The universal repetition of forms in nature somehow helps us interpret the SEM's new and obscure imagery. To many people SEM images such as the salmon fry's yolk sac in Plate 2.22 look like the surface of the moon (appropriately enough, *both* familiar and obscure!), while others may find that fossilized enamel from the microscopic tooth of a hand-sized dinosaur, as shown in Plate 2.23, looks like an aerial view of Kansas. In a similar way, although an SEM picture might be pleasing in the abstract, it captivates us even more, and we are relieved as well as surprised or amazed when we learn what it is.

Even scientists sometimes have trouble interpreting electron micrographs. If you're a researcher investigating something on the ultramicroscopic scale for the first time, how do you know what it's supposed to look like? Are the features real or are they artifacts created by the sample preparation process? Are they real, but not what you think they are? Is the appearance created from choosing one way to use the microscope over another? (What happens, for instance, if you change the voltage?) These questions always need to be addressed before a scientific project is valid. To add to the confusion, unrelated features in the same sample can resemble each other, making it difficult to determine which is which. (The feature in Plate 2.29 is a good example.) Since most of the microworld is still unexplored, SEM investigations can lead us into new areas of research by revealing structures that imply hitherto unsuspected natural processes, such as the forces acting on the Antarctic sand grain in Plate 3.22. With experience, however, pioneer research efforts eventually become well-charted territory. New frontiers yield to exploration and a consequent increase in our understanding of another small segment of the universe.

So it seems that in making use of the scanning electron microscope, a tool intended for investigative and descriptive purposes, the very act of liberating something familiar from its workaday boundaries, or of revealing aspects of something more exotic, can create riddles and solve them, please the eye, and send the imagination soaring. Perhaps the micrographs in this book will affirm that in the routine technical pursuits of science we often find an unexpected elegance.

1.1

1 · TUTORIAL

MAGNIFYING A FEATHER

THIS CHAPTER OFFERS A BRIEF INTRODUCTION TO THE MICROWORLD, by using a familiar object to demonstrate the mechanics of magnification and the look of scanning electron imagery.

Plates 1.2 and 1.3 show a small goose feather that was prepared for examination in the SEM. The feather was mounted with double-sided tape onto a half-inch diameter aluminum "stub," which was then placed into a vacuum chamber and coated with an alloy of gold and palladium in a low pressure atmosphere of argon gas. Conductive metal coatings (or carbon, in the case of samples intended for chemical analysis while they're being viewed) facilitate the formation of pictures by grounding the sample, which allows the otherwise chaotic electrons to be controlled during the imaging process. The photograph was taken with an ordinary camera and is printed here in both the actual size of the sample and at twice its actual size (that is, a magnification of 2).

The series continues with scanning electron micrographs of the same feather, demonstrating what it looks like at successively higher magnifications. Each picture is contained somewhere within its predecessor and shows a proportionately smaller part of the feather as the magnification increases. Feathers are believed to have evolved from the scales of a small, carniverous dinosaur; here we can see that their intricate parts interlock for aerodynamic efficiency.

Magnifications throughout the book are indicated by numbers following the symbol "x." For example, "x 100" means that the area in the picture is shown 100 times its actual size.

Plate 1.2 Goose feather, x 1

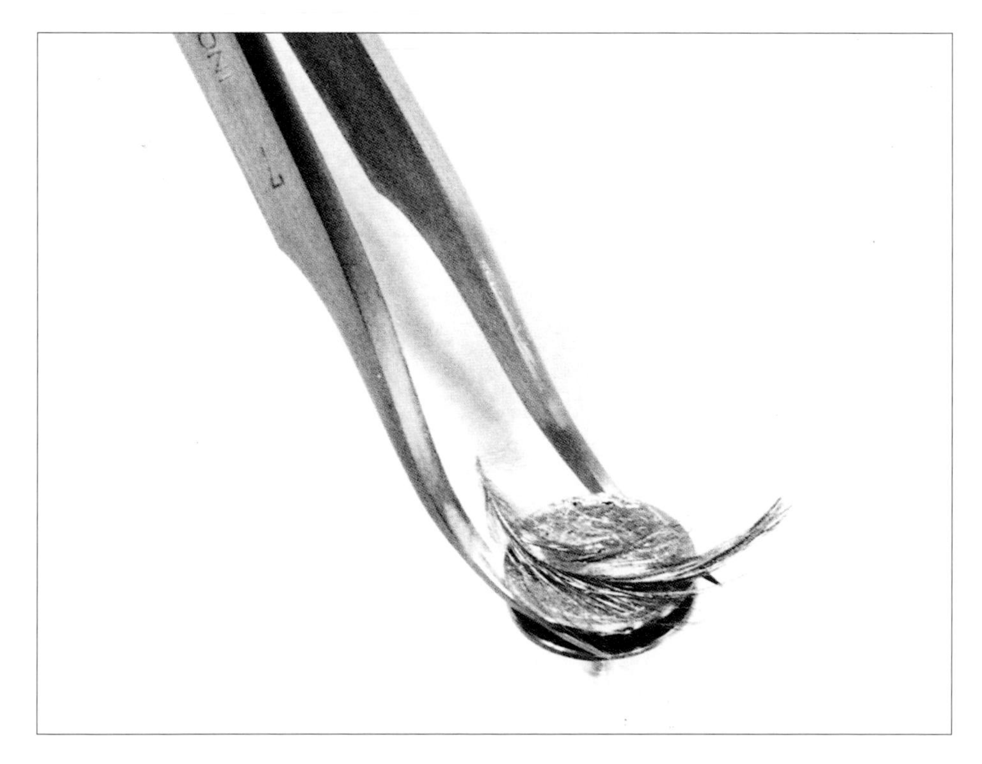

Plate 1.3 Goose feather, x 2

1.4 x 20

1.5 x 40

1.6 x 120

1.7 x 265

1.8 x 635

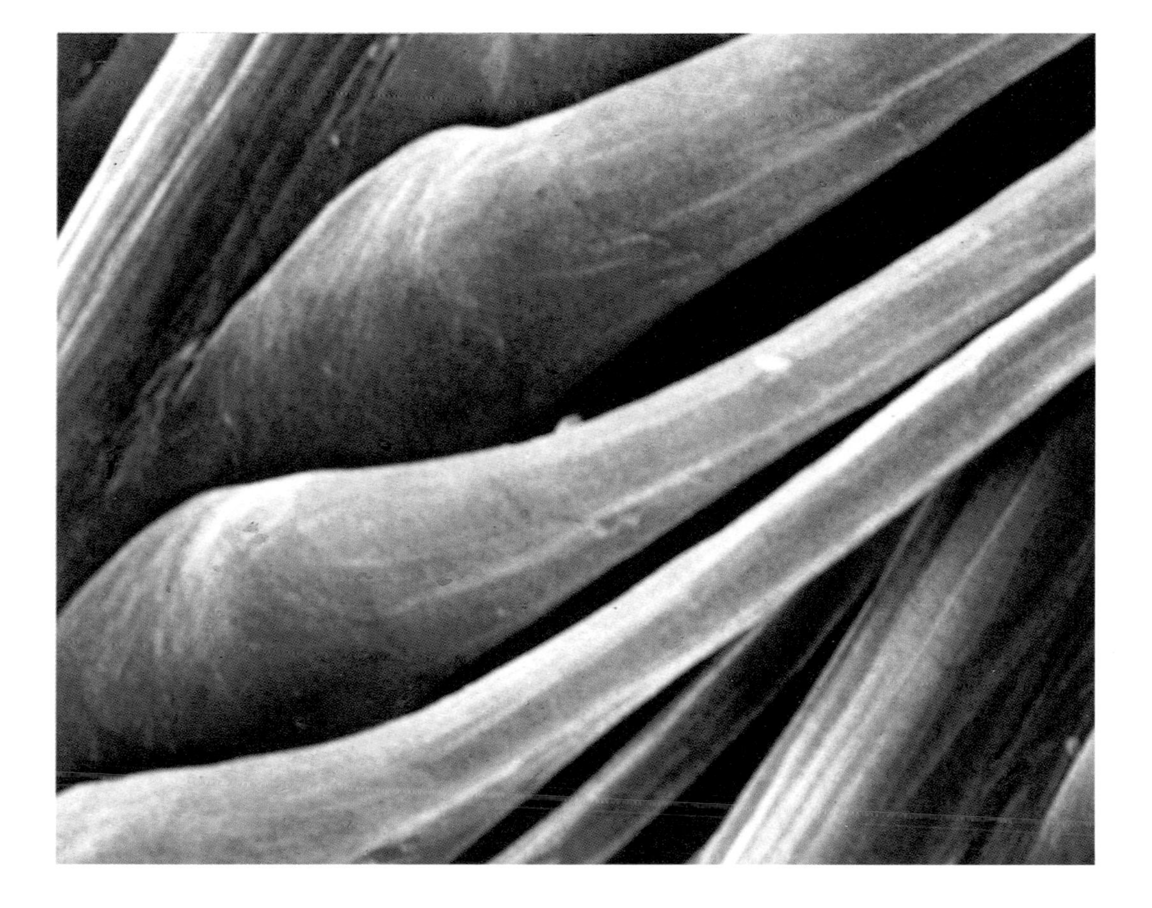

1.9 x 1,560

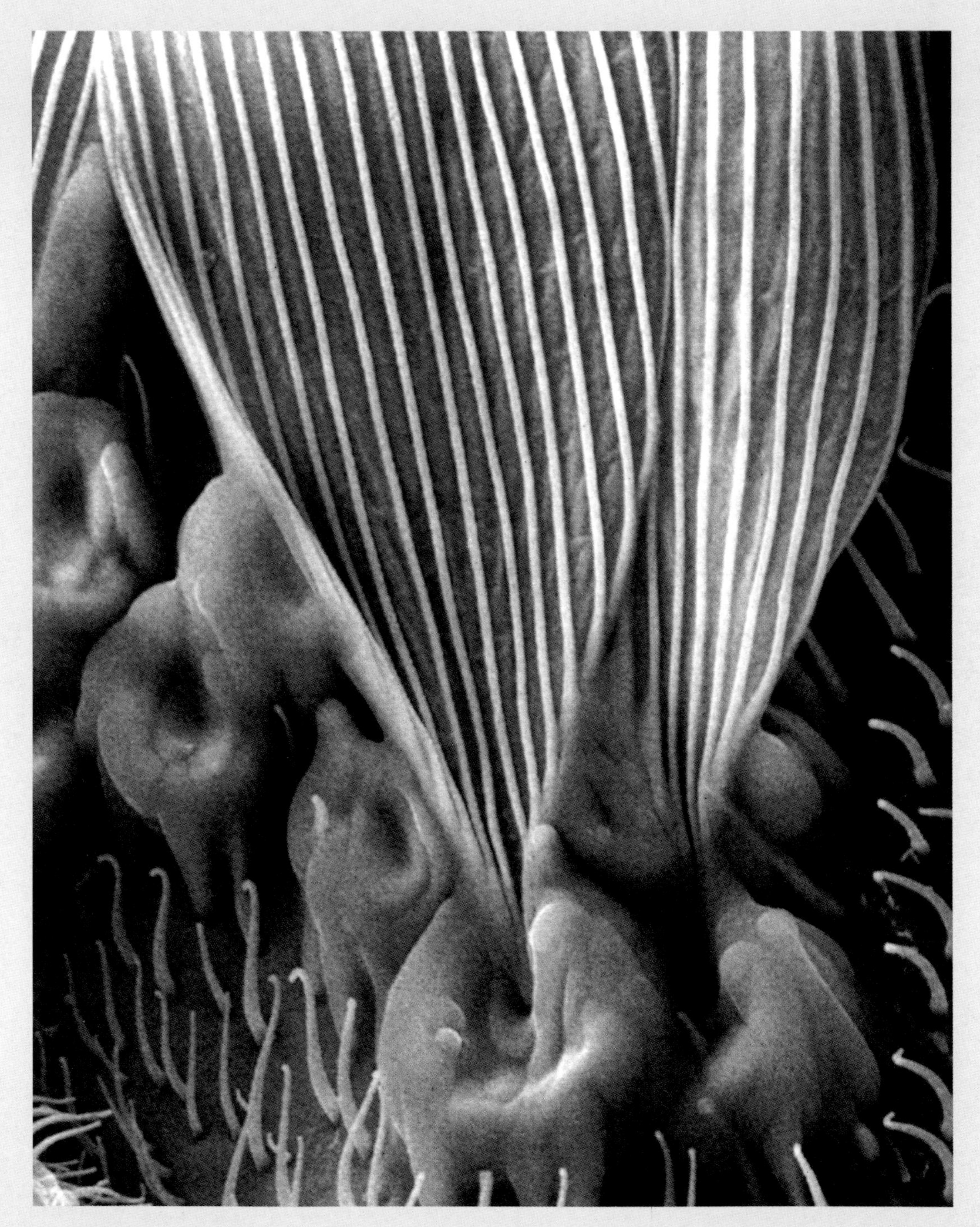

2.1

2 · ABSTRACTS

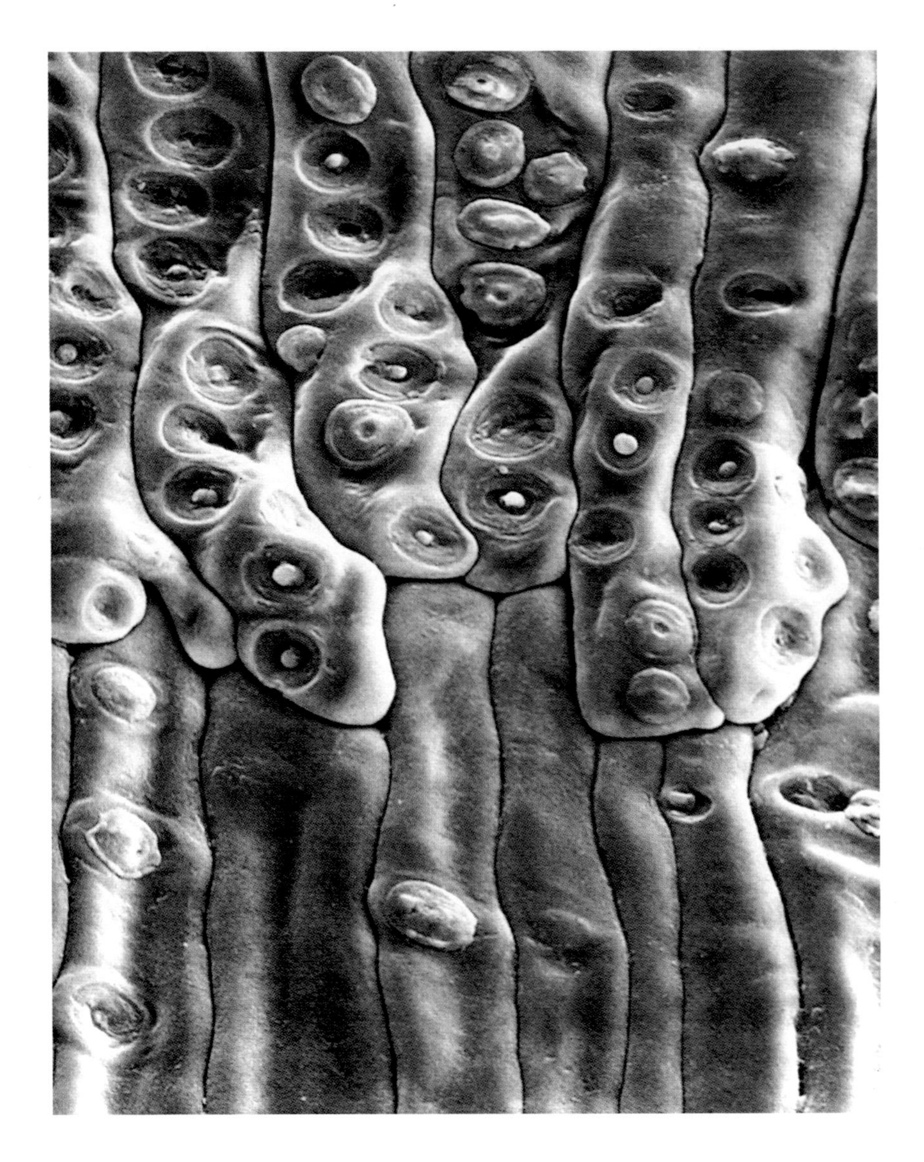

2.2 TWO SETS OF CELLS ON AN ANCIENT CONIFER: petrified wood from Tierra del Fuego, x 600 (7)

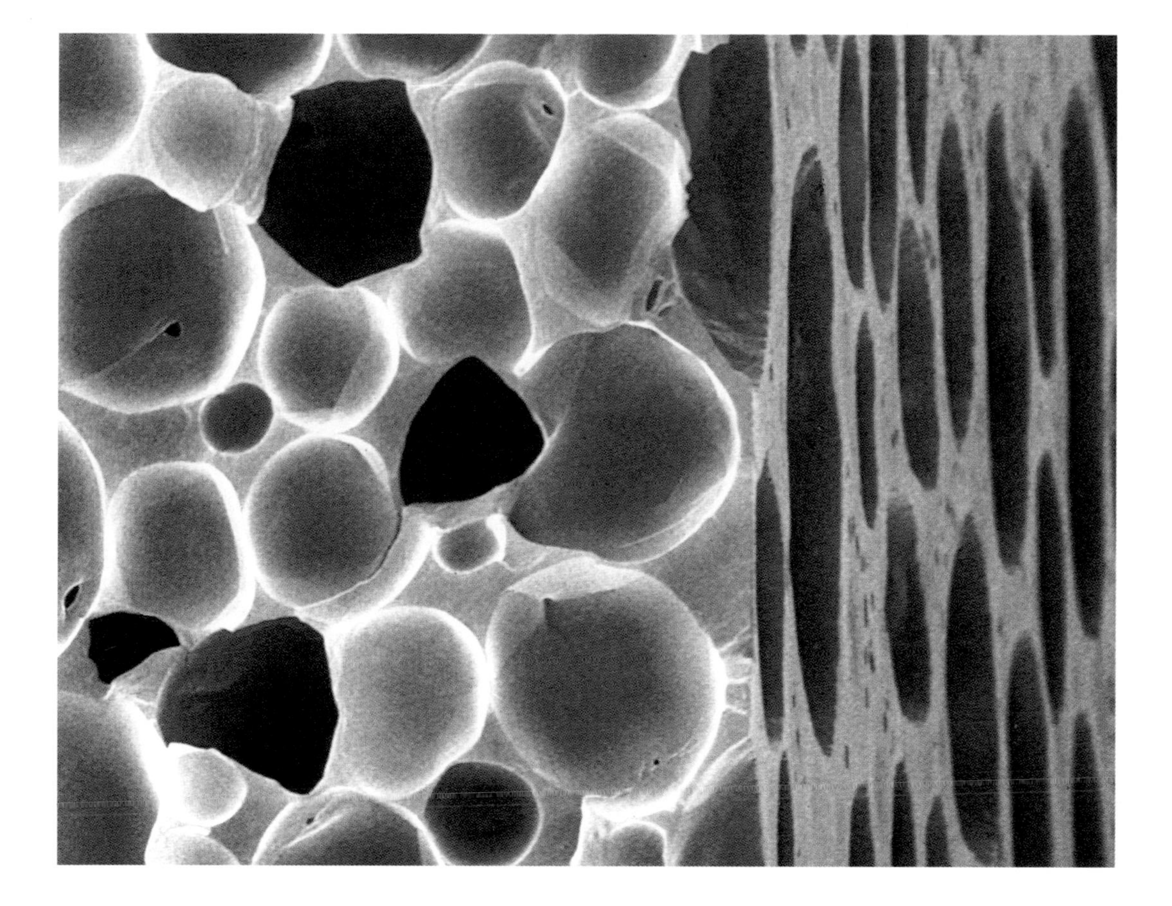

2.3 INDUSTRIAL FILTER: at the edge between its cross-section on the left, and one of the sides falling away to the right, x 1,300 (8)

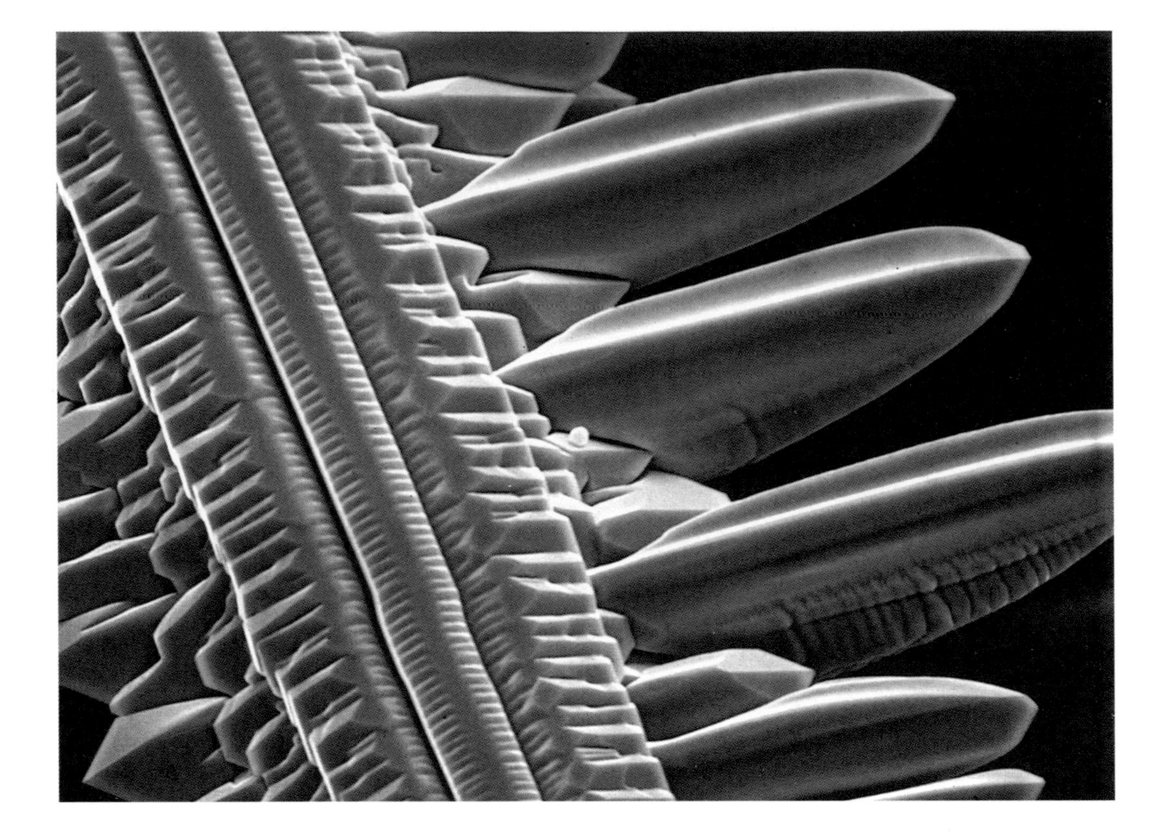

2.4 Surface detail on a tungsten wire that was heated until it flamed, x 2,100 (9)

2.5 Betsy Beetle, Scene 1, x 450 (6)

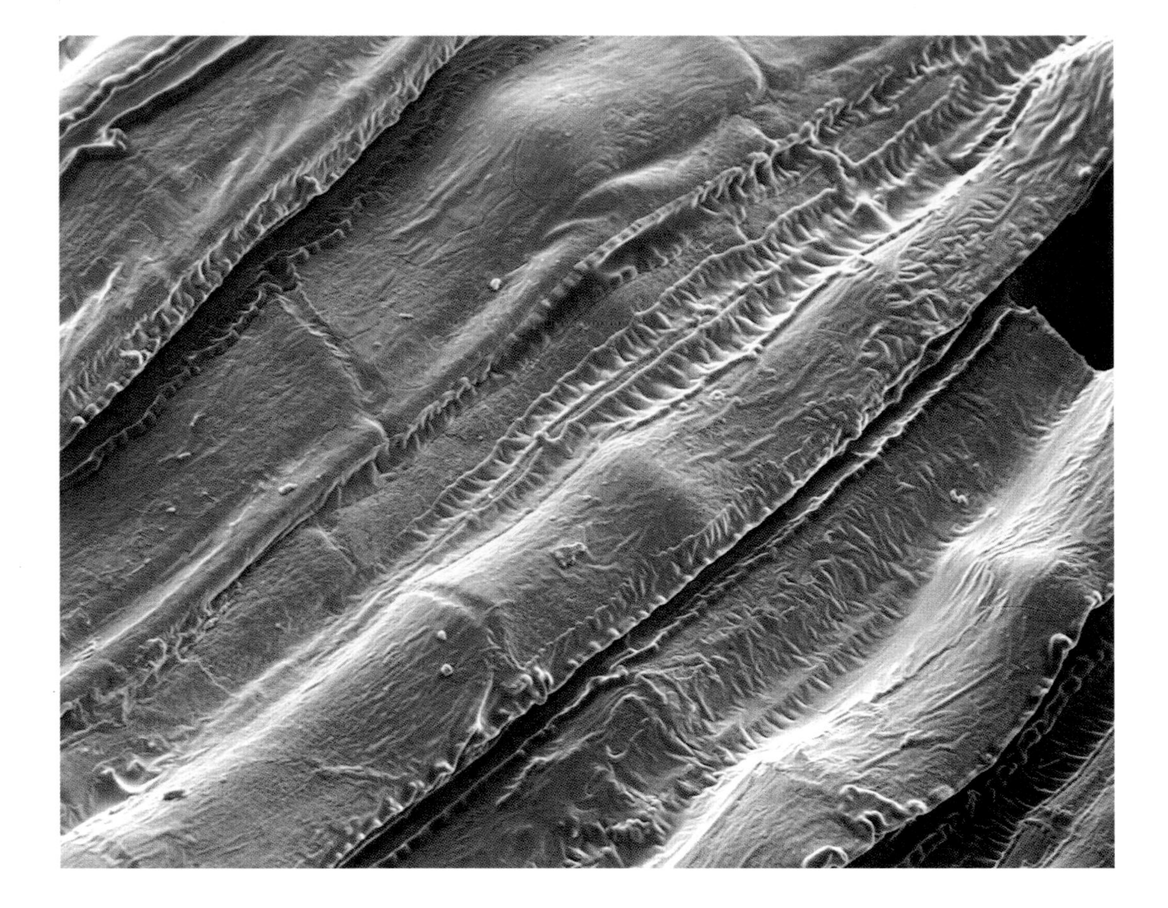

2.6 LEFTOVERS: on the fin of a smoked whitefish served by Aunt Betty, x 85 (10)

2.7 Serrated edge of a tooth from a tiny carnivorous dinosaur, x 100 (11)

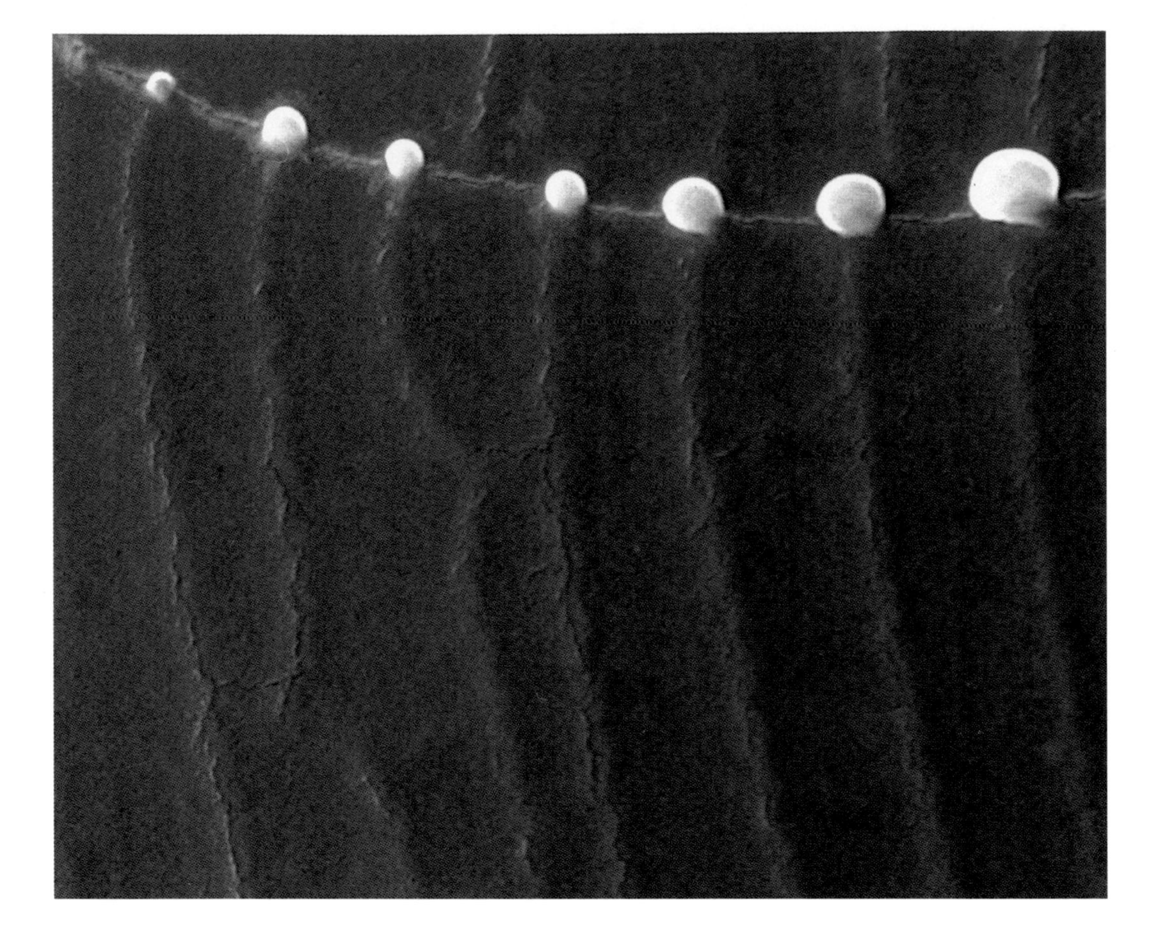

2.8 Circulatory system in the fin of a salmon fry, x 2,600 (12)

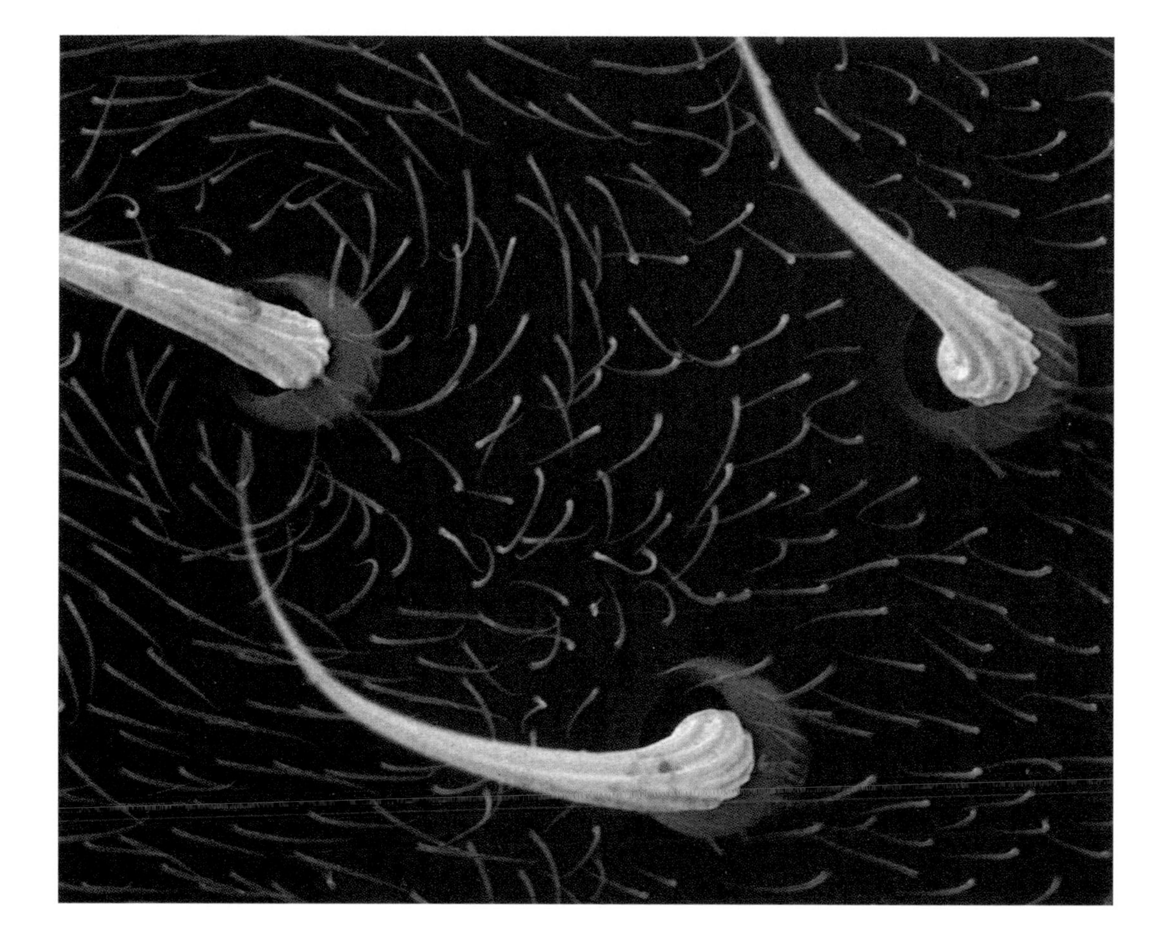

2.9 Surface detail on an anonymous bug that wandered into the home of the microscopist and was prevailed upon to donate itself to Science, x 1,100

2.10 Franklin micromineral, x 700 (13)

2.11 At the edge of the wing of a Painted Lady butterfly, x 400 (6)

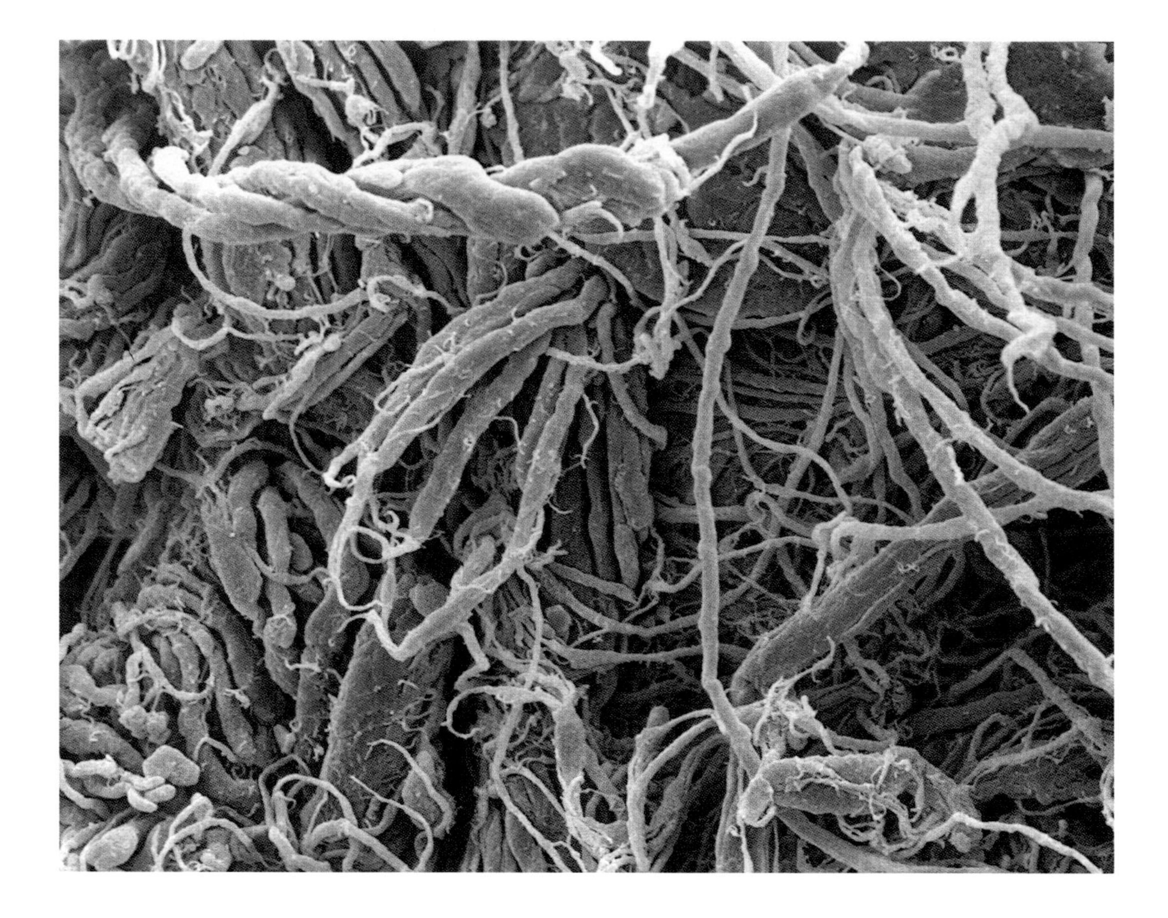

2.12 Suede, x 450 (14)

2.13 CHRYSOTILE: the mineral source of asbestos, x 7,300 (15)

2.14 Surface detail, burned tungsten wire, x 900 (9)

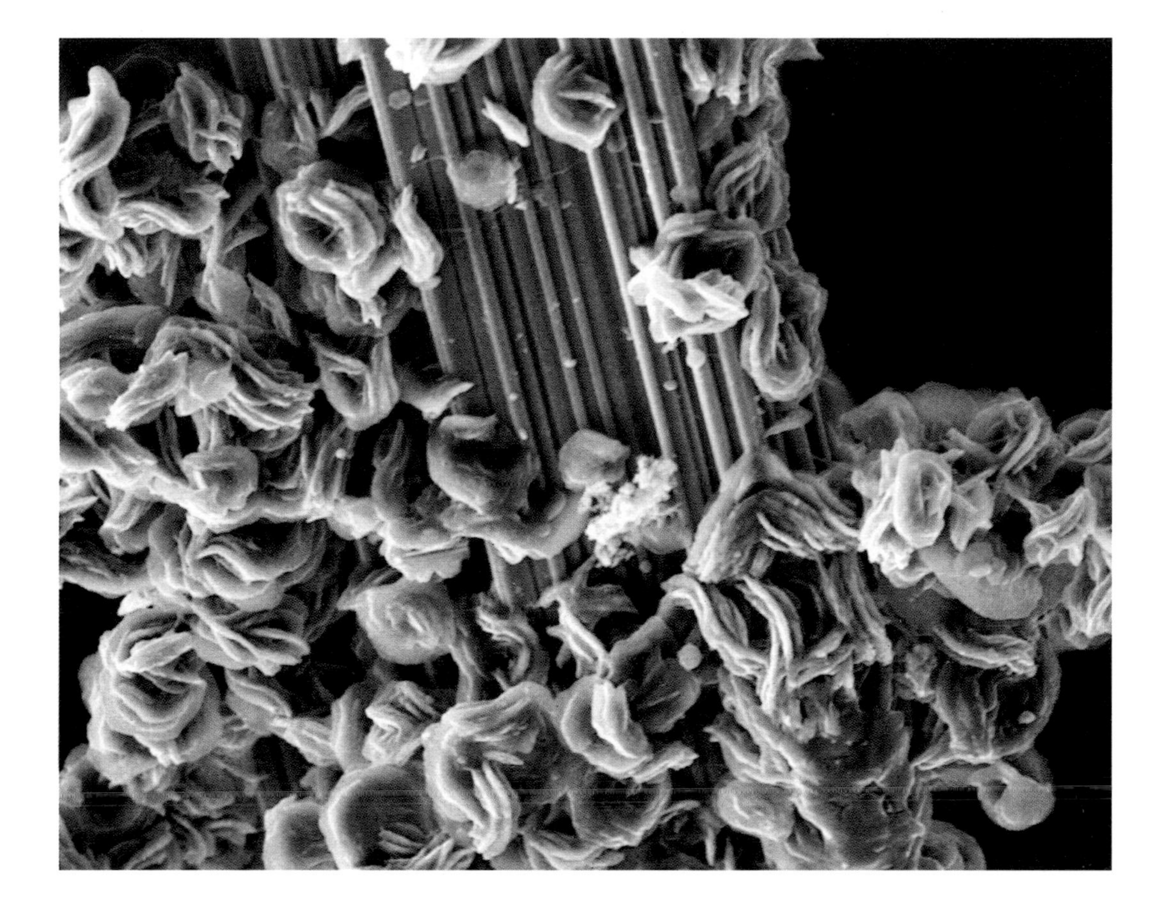

2.15 Franklin micromineral, x 1,700 (13)

2.16 Surface detail on a fragment of the oldest known rock on Earth, x 13,700 (16)

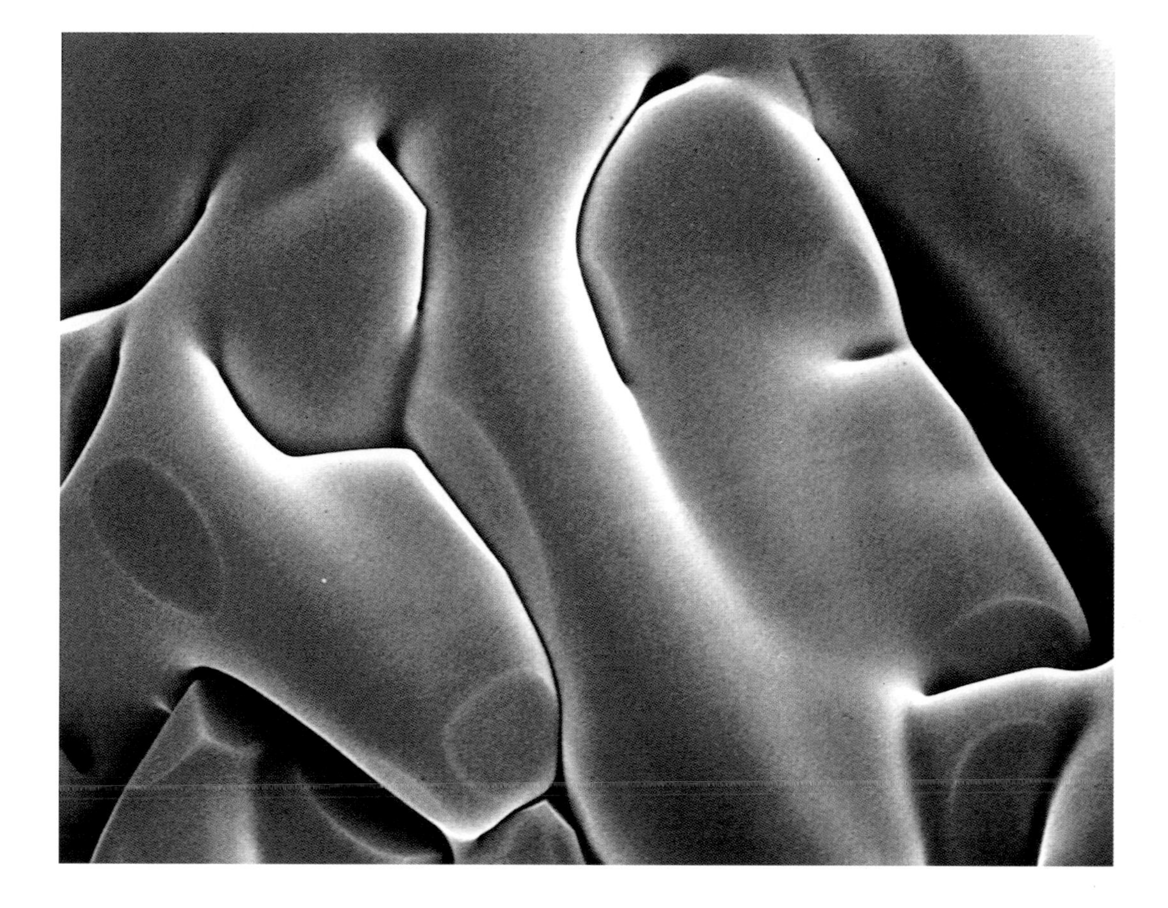

2.17 Surface detail on burned tungsten wire, x 2,800 (9)

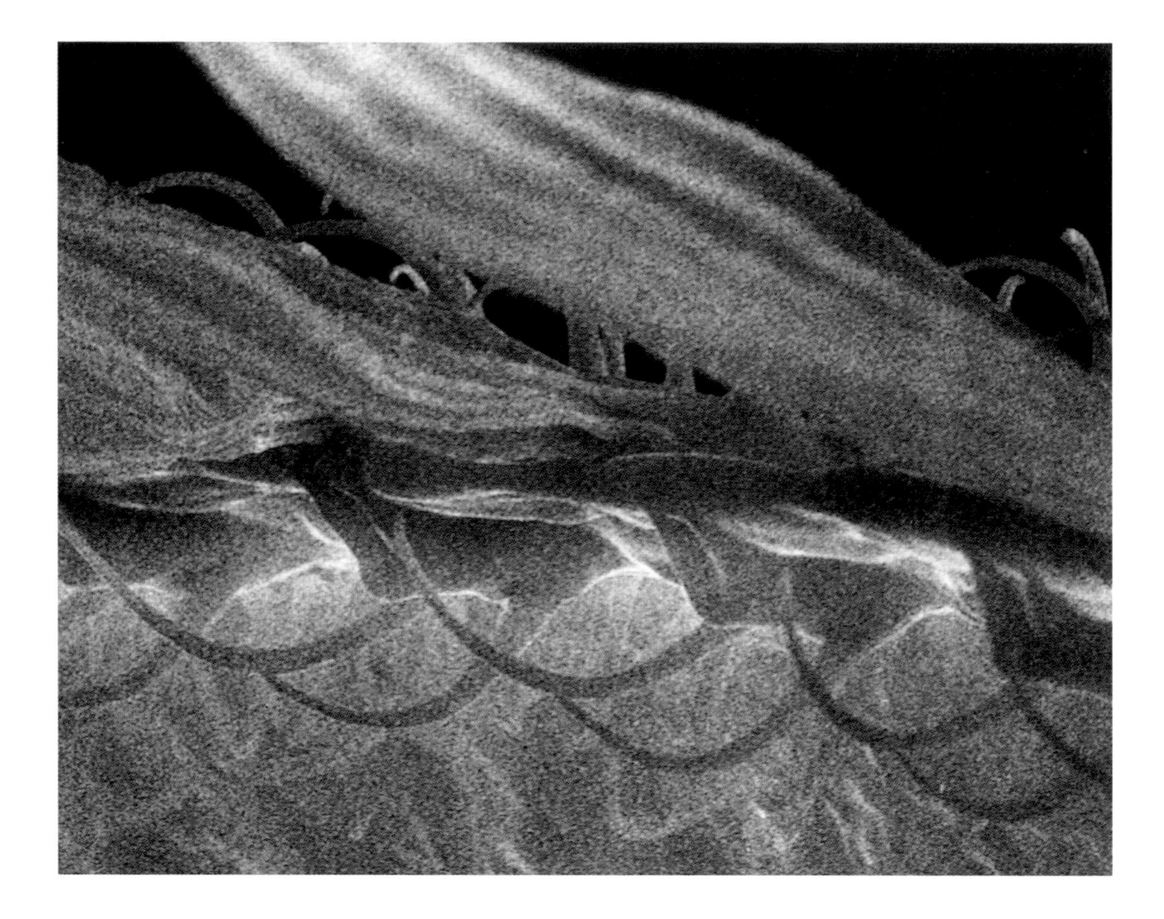

2.18 MOSQUITO WING SERIES: electron backlighting at the edge of the wing, x 4,300 (6)

2.19 GEORGIA O'KEEFFE REVISITED: tropical coccolith, x 20,800 (3, 17)

2.20 The effect of vacuums and hot electrons on double-stick tape, x 250

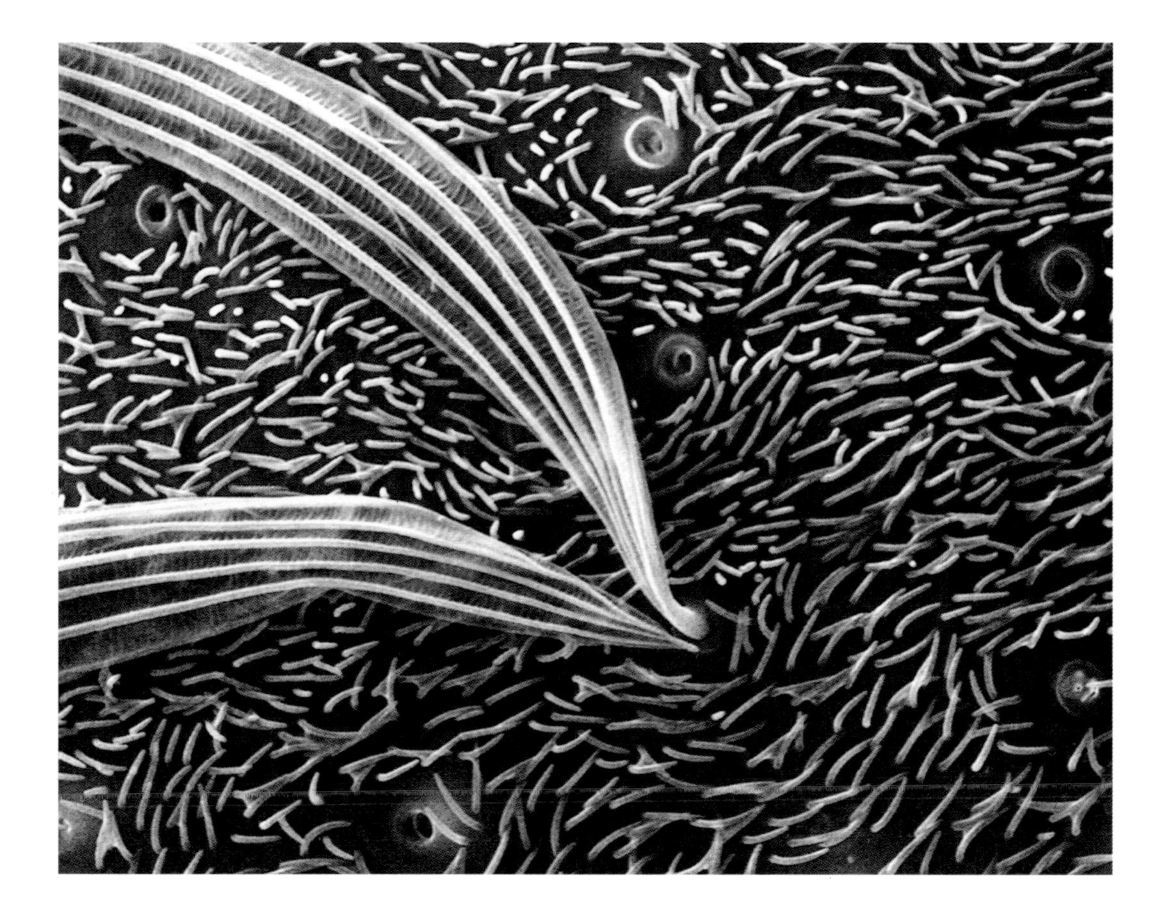

2.21 MOSQUITO WING SERIES: surface detail, x 2,300 (6)

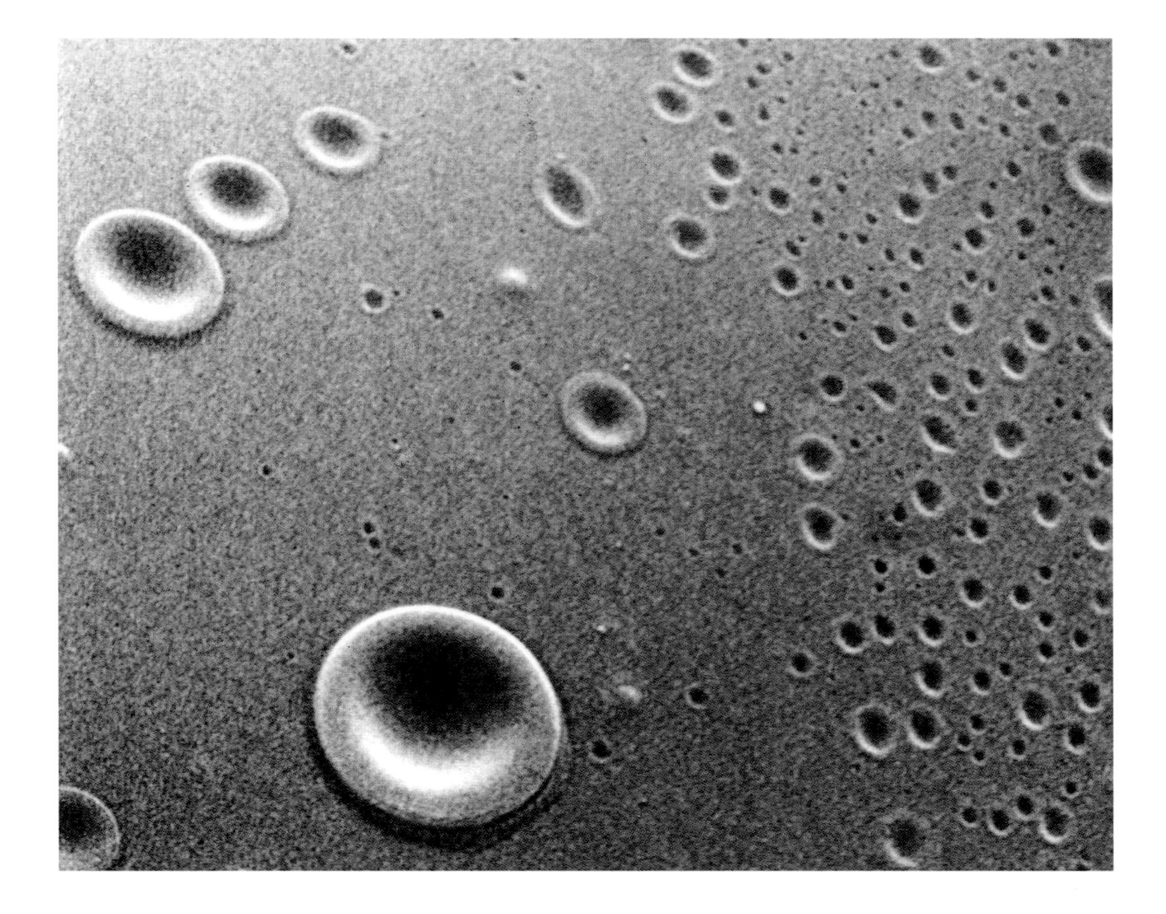

2.22 Structures on the yolk sac of a three-week-old salmon fry, x 1,200 (12)

2.23 Surface detail of a microscopic tooth from a very small dinosaur, x 200 (11)

2.24 Minerals from Zabargad, an island in the Red Sea that was once a part of the Earth's upper mantle, x 2,200 (18)

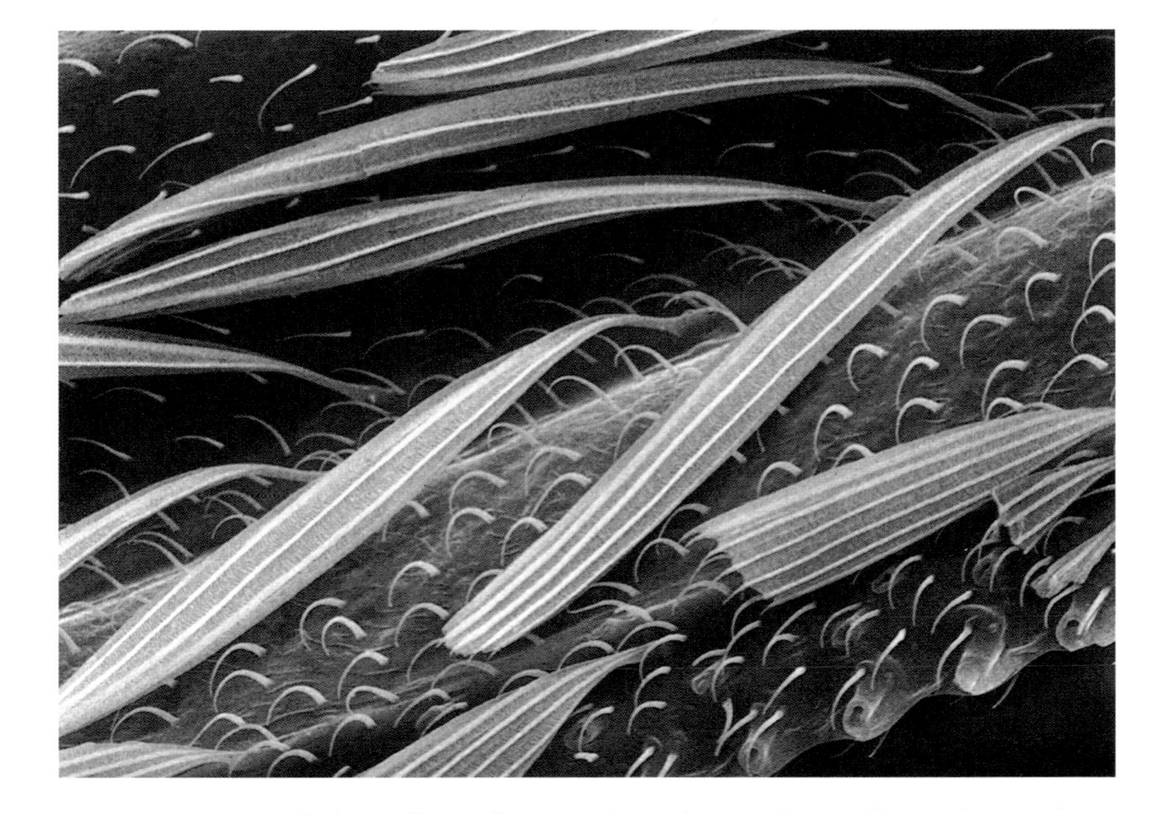

2.25 MOSQUITO WING SERIES: just inside the edge, x 1,000 (6)

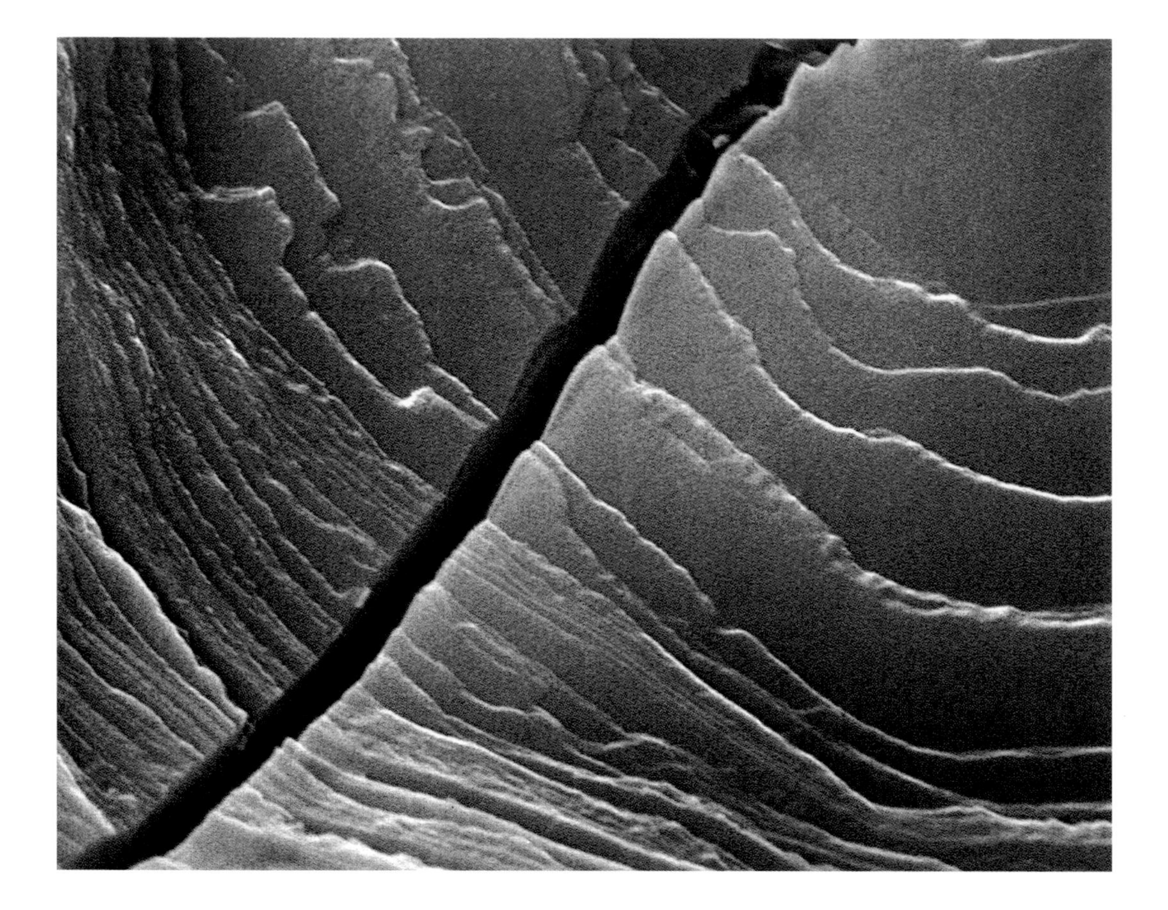

2.26 A tiny crack in a fragment of rock nearly four billion years old, x 14,200 (16)

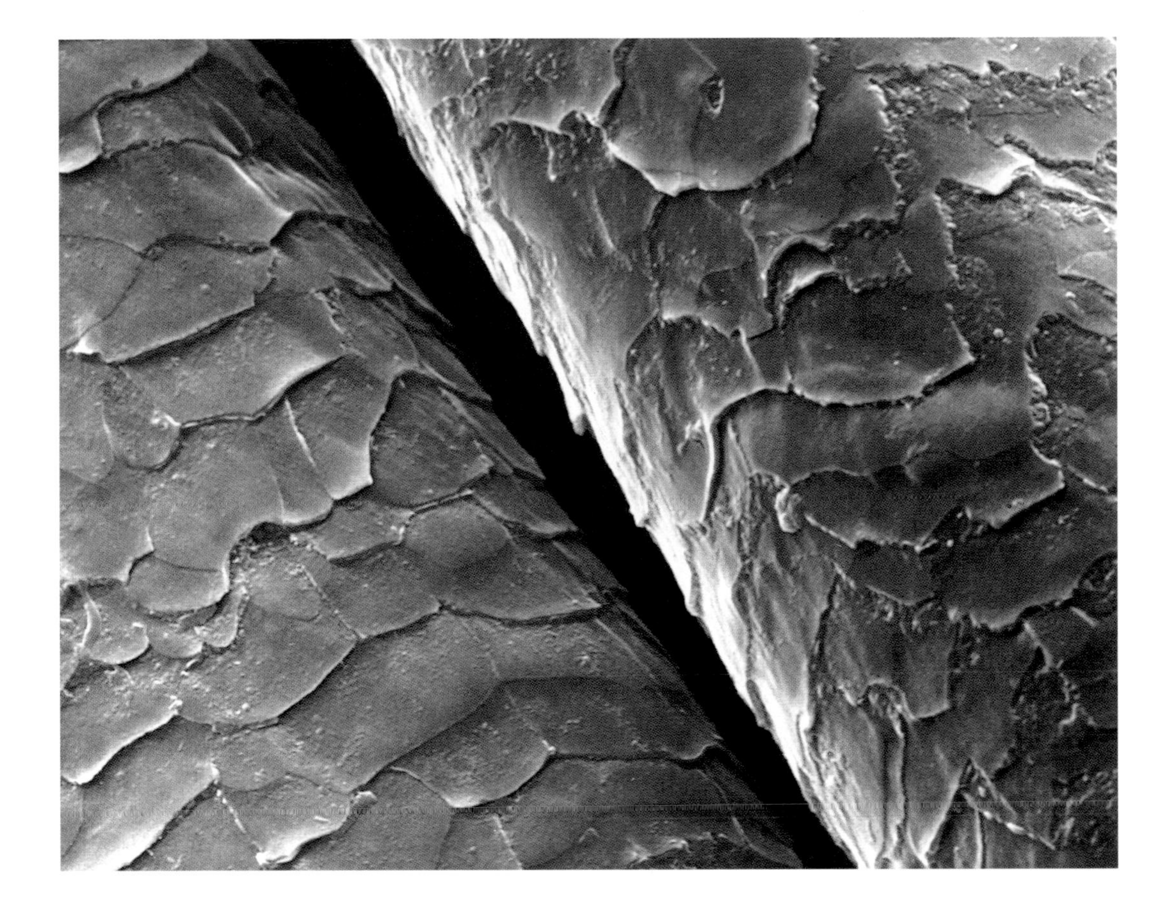

2.27 KATHY'S HAIR: two strands lying side by side, x 1,600 (19)

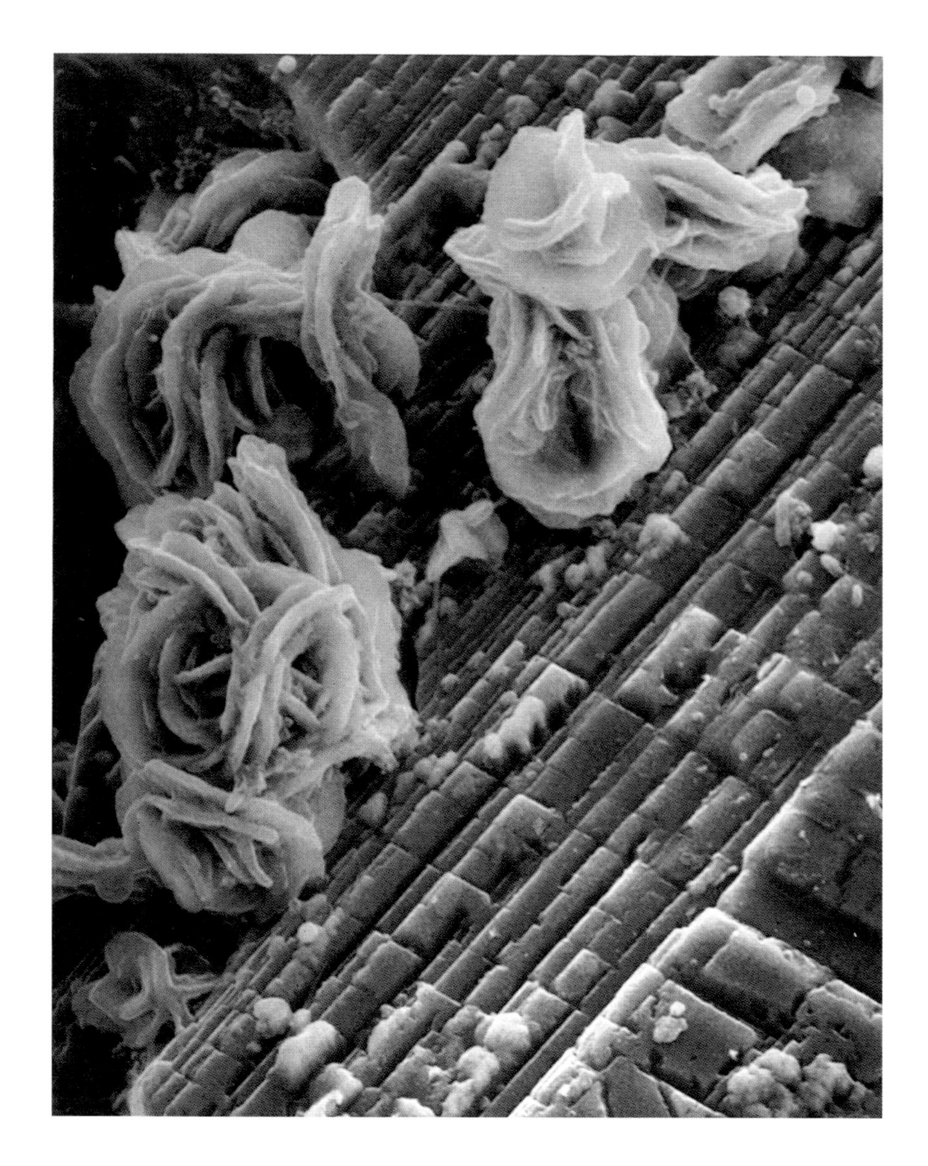

2.28 Franklin micromineral, x 1,900 (13)

2.29 Bacteria (or cholesterol?) on human heart muscle, x 11,200 (20)

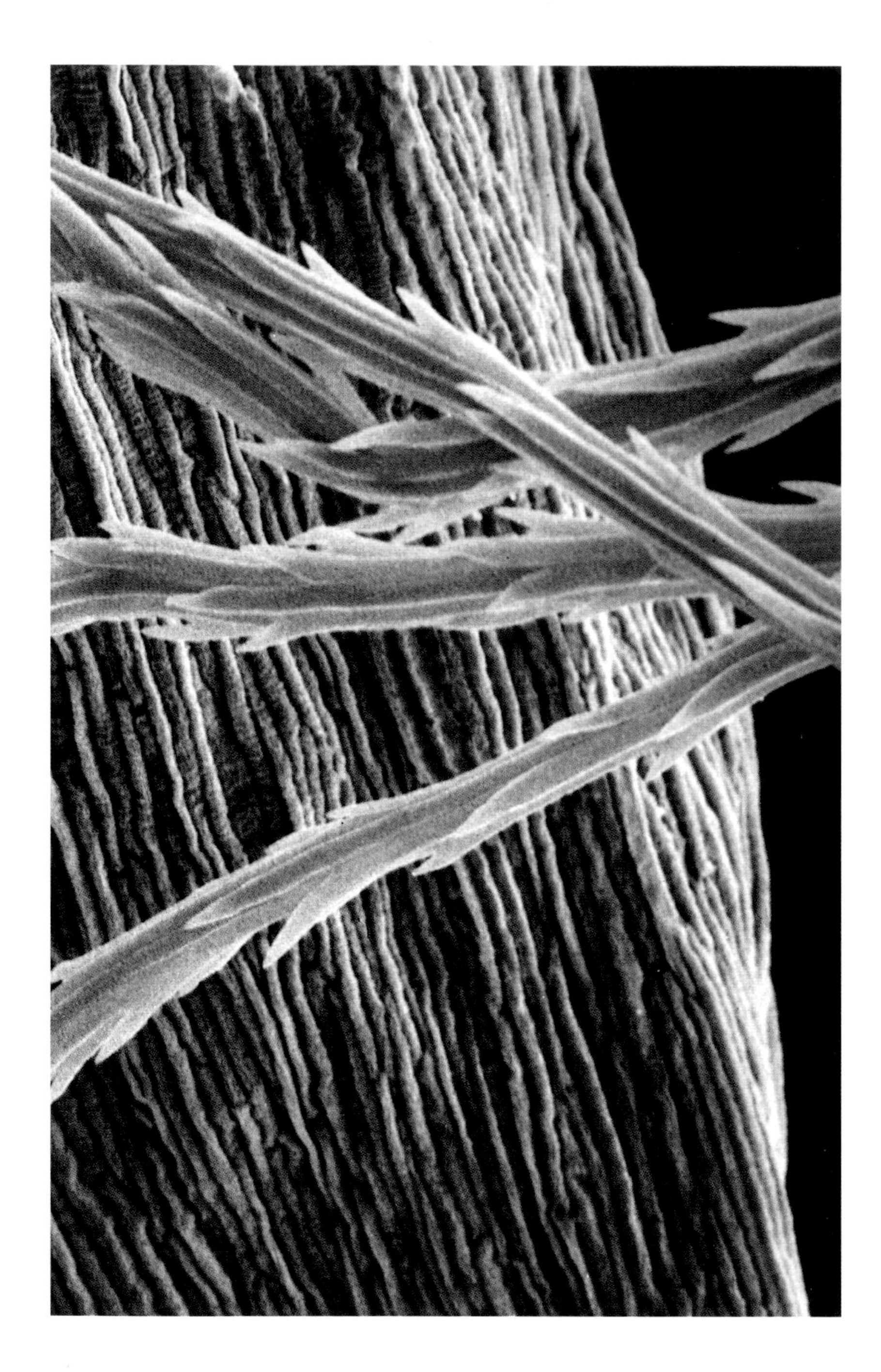

2.30 Random bits of a goldenrod plant, x 400

2.31 MOSQUITO WING SERIES: at the edge, x 800 (6)

2.32 Plumose antenna of a male mosquito, x 700 (6)

2.33 ORDER IN CHAOS: fragment of an Arctic diatom, x 1,400 (2, 21)

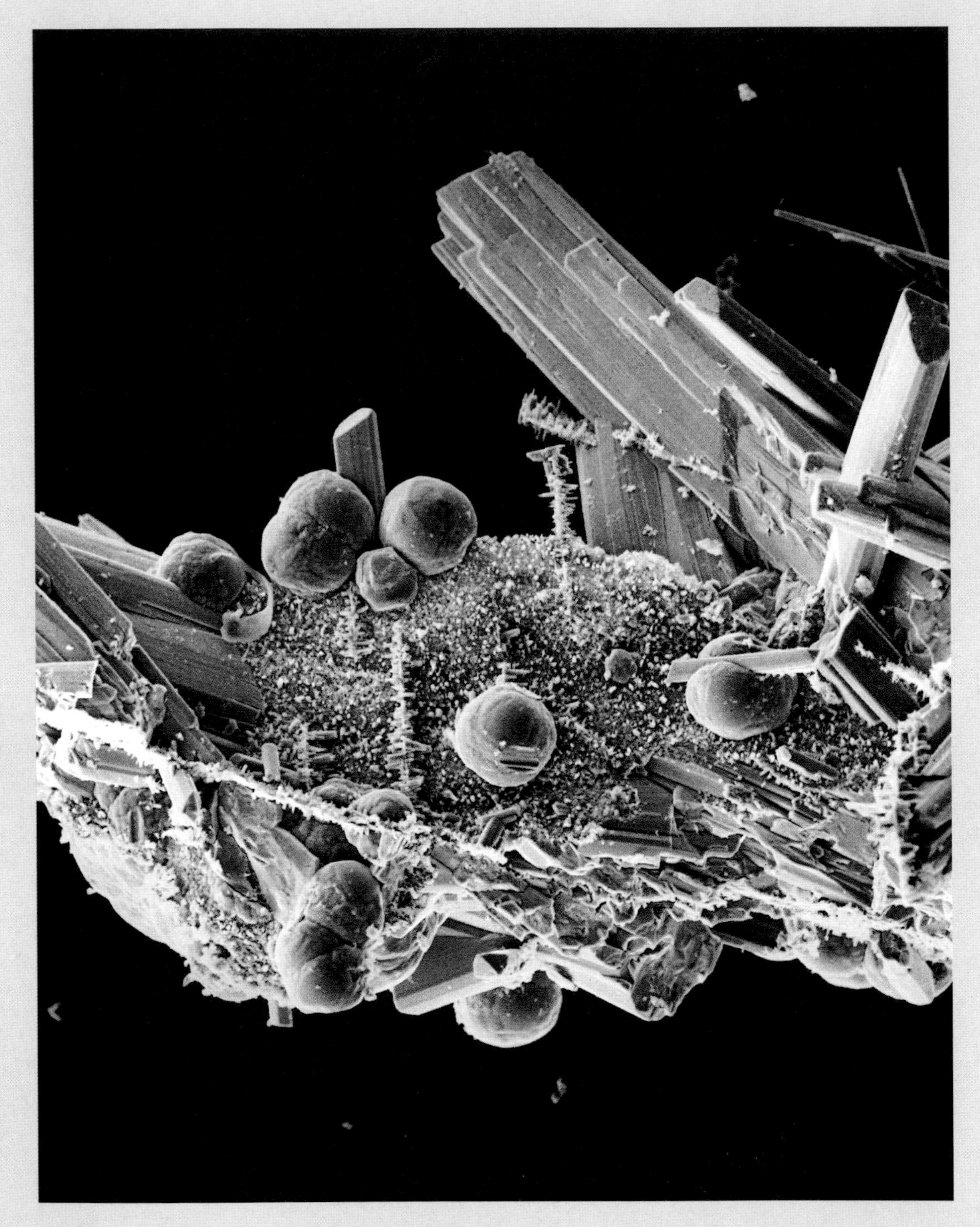

3.1

3 · MICROSCAPES

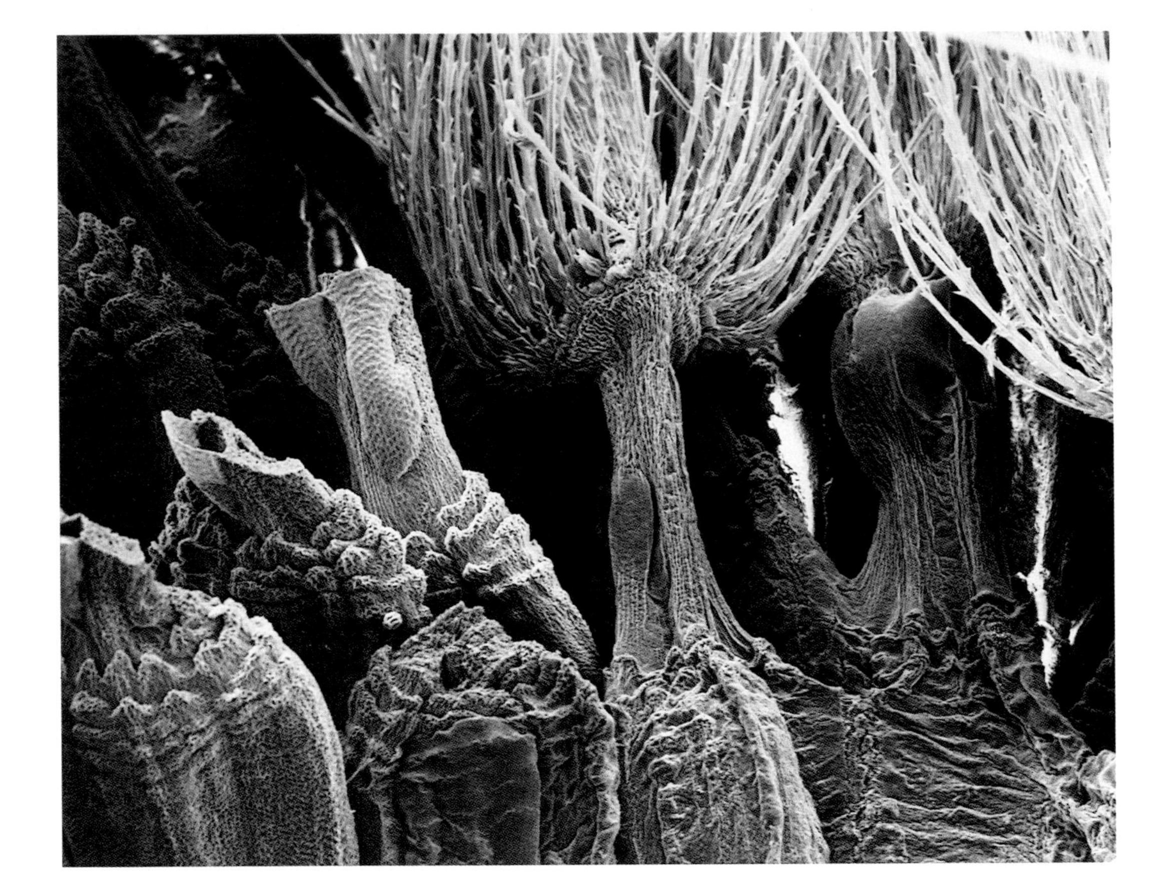

3.2 A row of dandelion seeds inside the blossom (some decapitated), x 75

3.3 MOSQUITO WING SERIES: at the edge, x 1,800 (6)

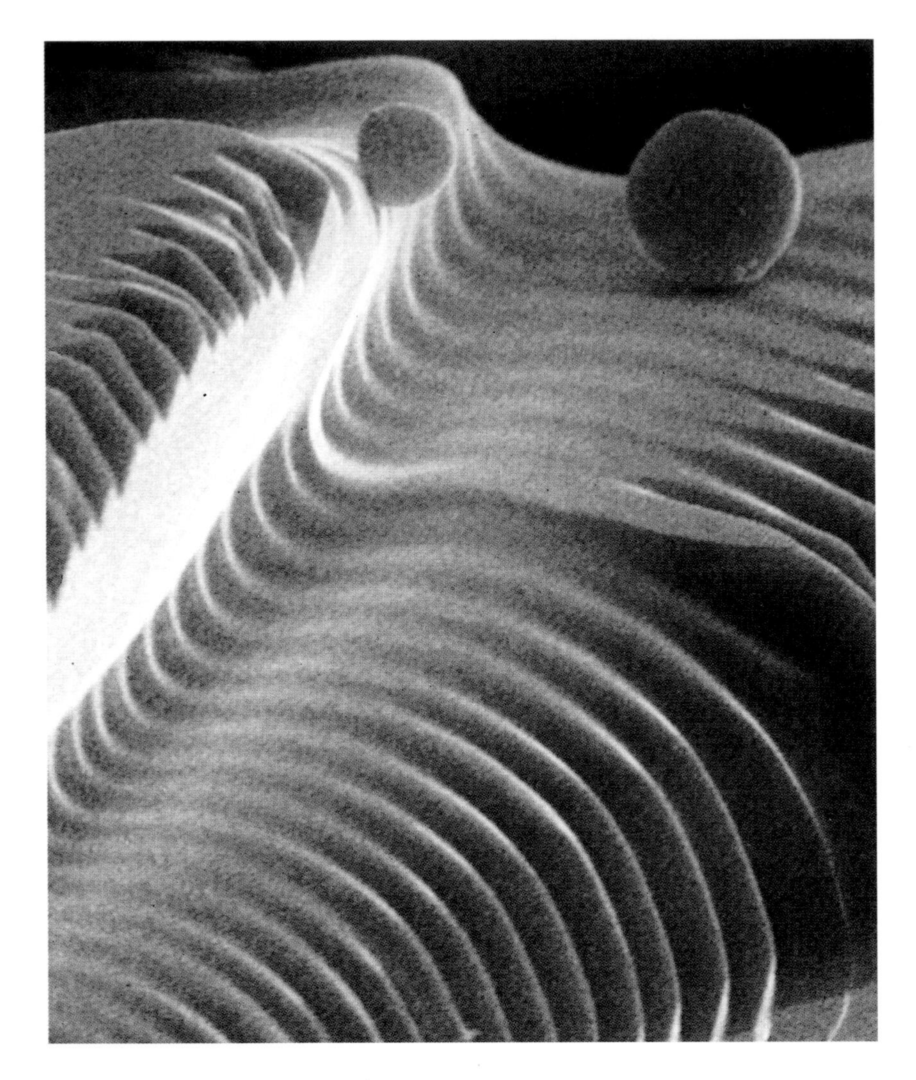

3.4 Surface detail on burned tungsten wire, x 10,500 (9)

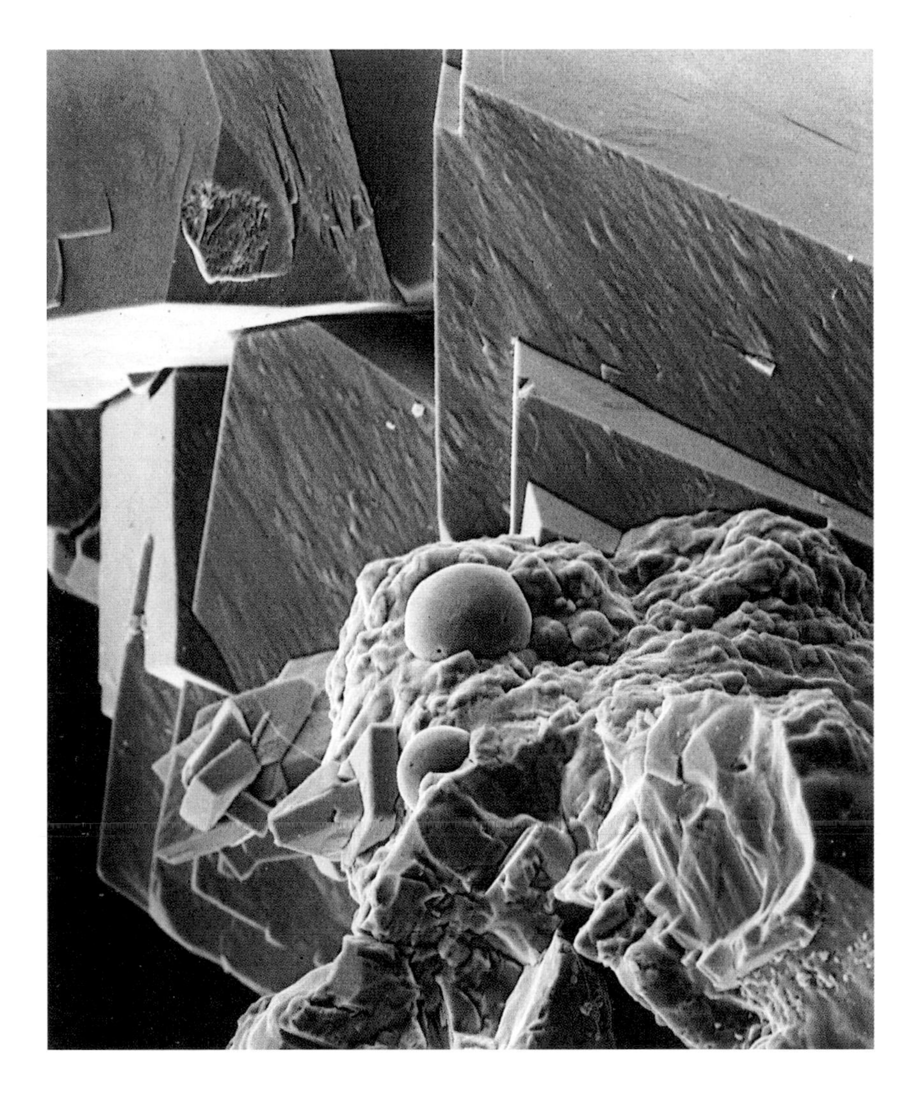

3.5 Franklin micromineral, x 375 (13)

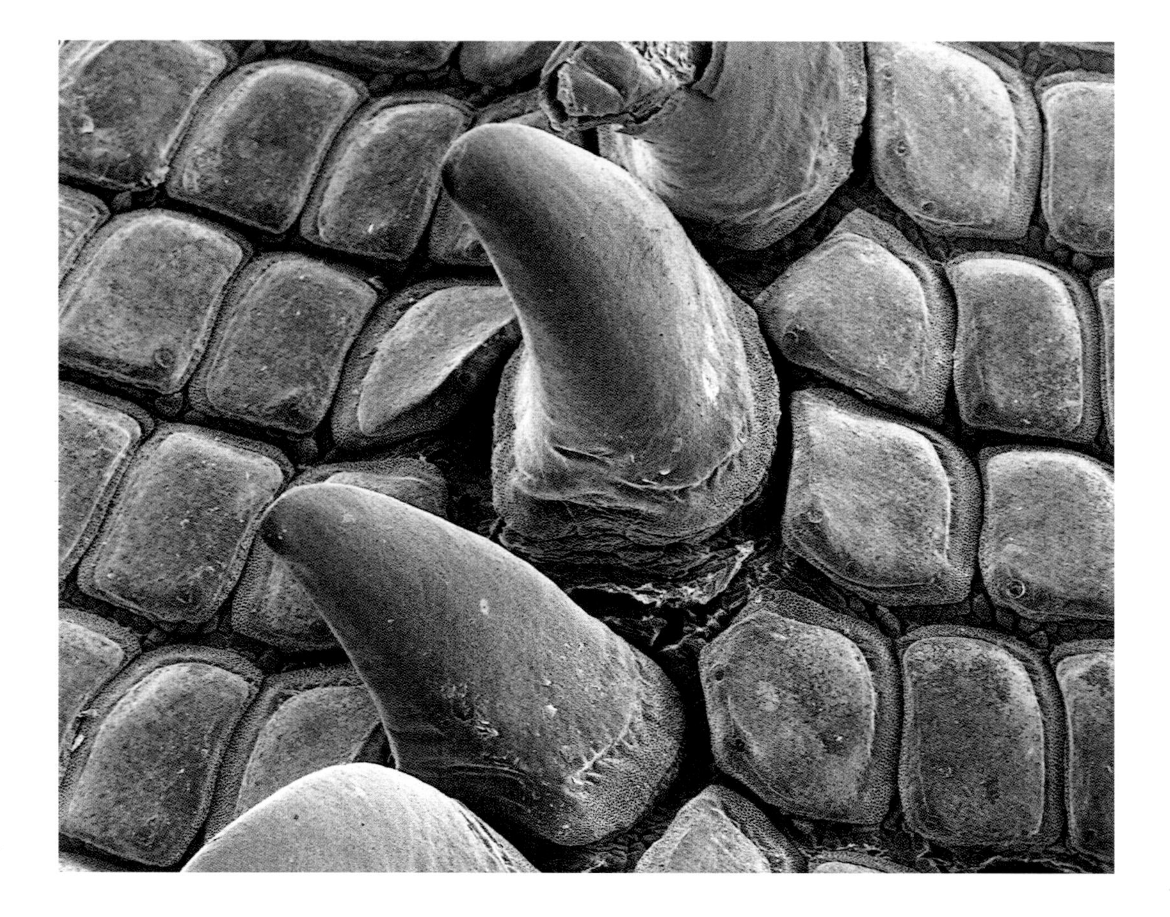

3.6 Spines on the tail of an iguana, x 20 (22)

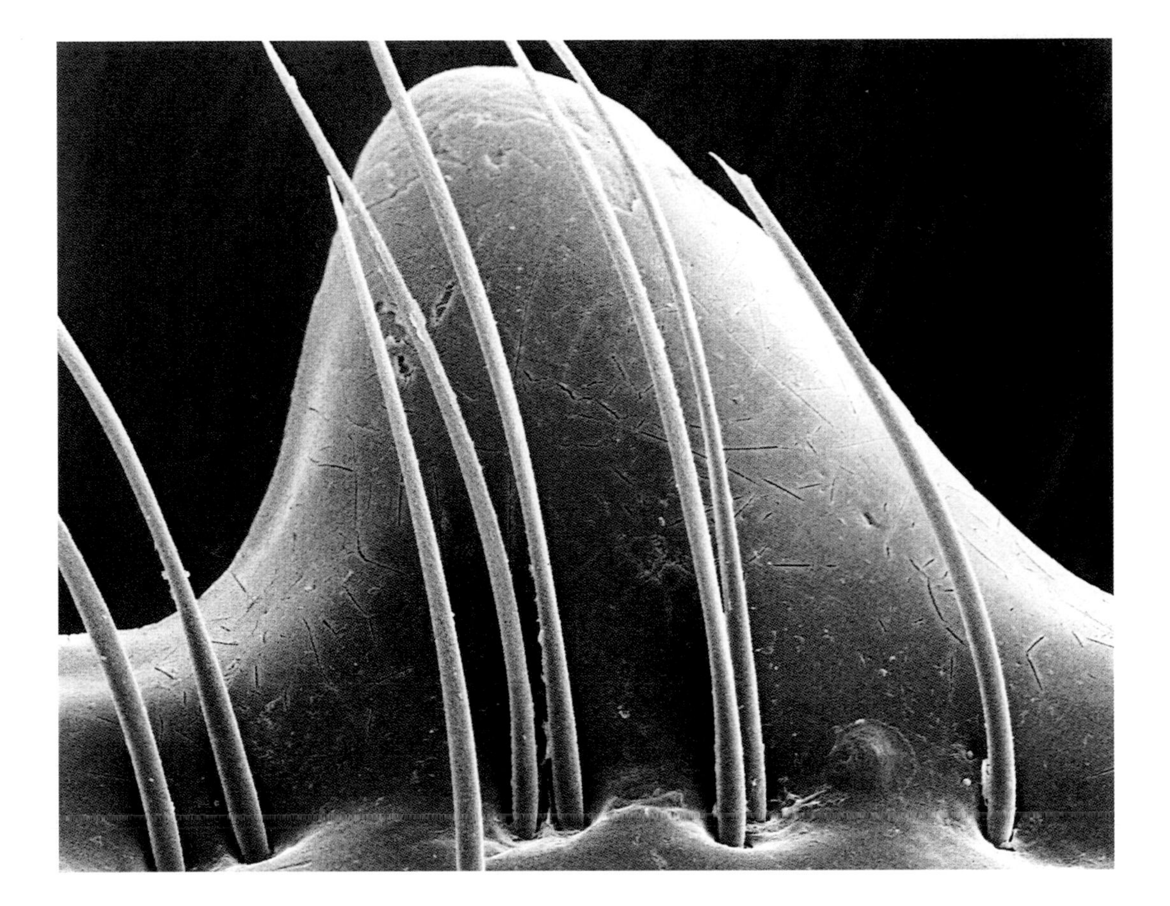

3.7 Betsy Beetle, Scene 2, x 250 (6)

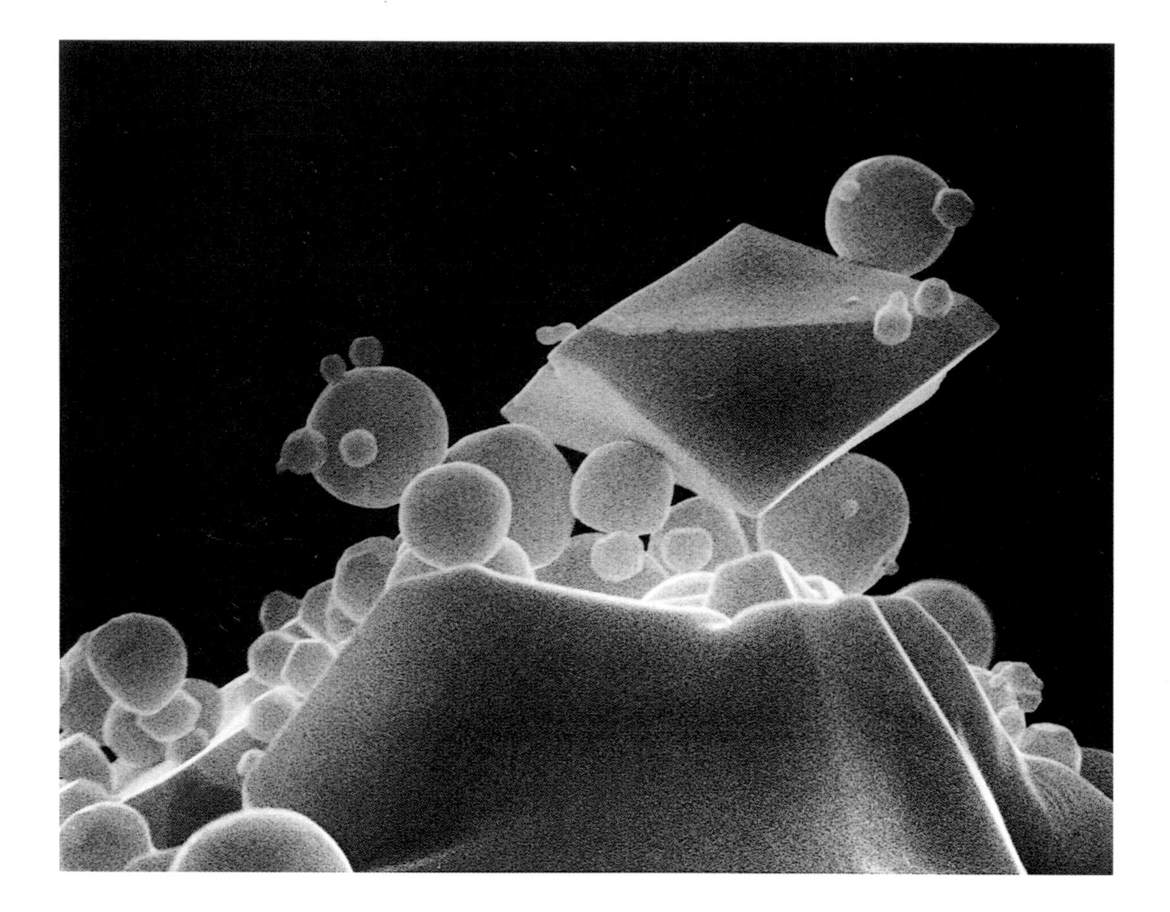

3.8 At the edge of a burned tungsten wire, x 18,500 (9)

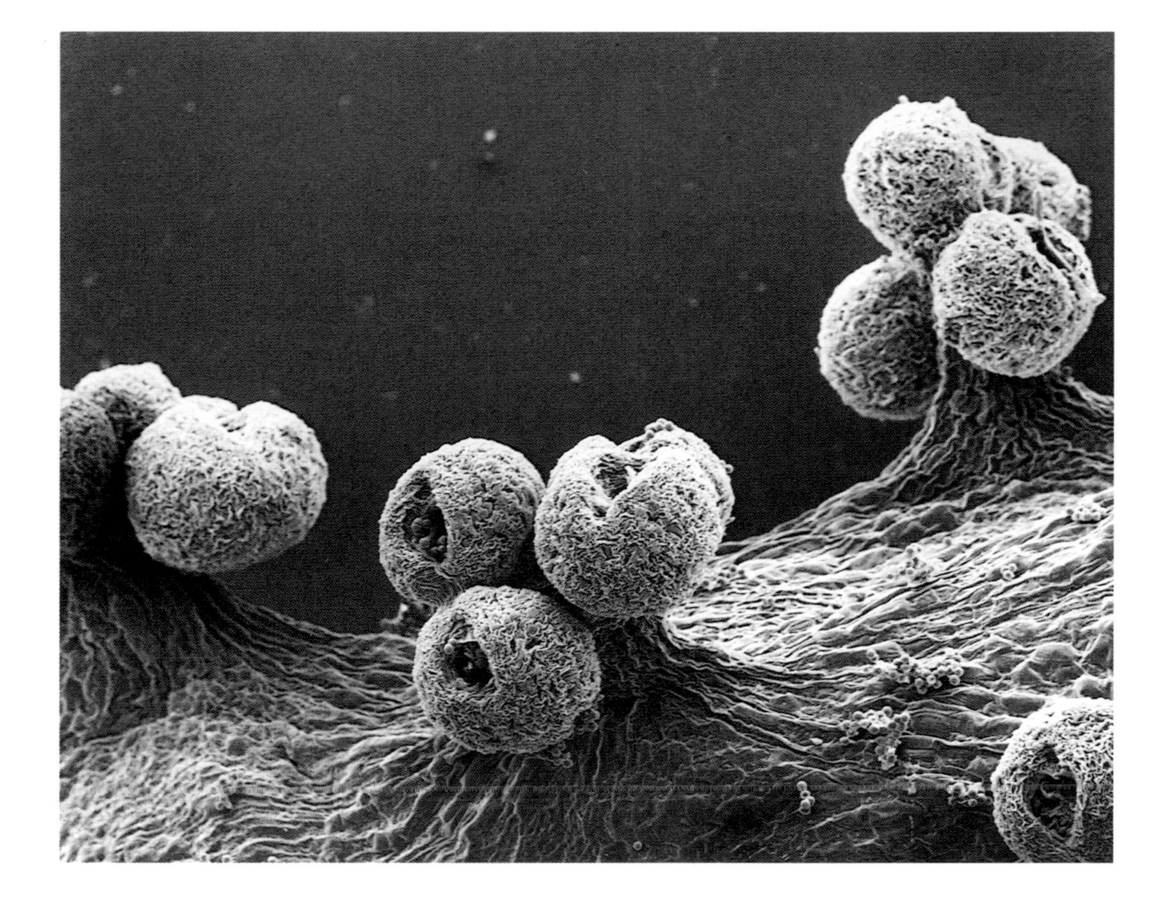

3.9 Pollen sacs on a Jack-in-the-Pulpit, with pollen inside them and scattered around, x 70 (23)

3.10 A tumble of coccoliths, x 10,600 (3, 17)

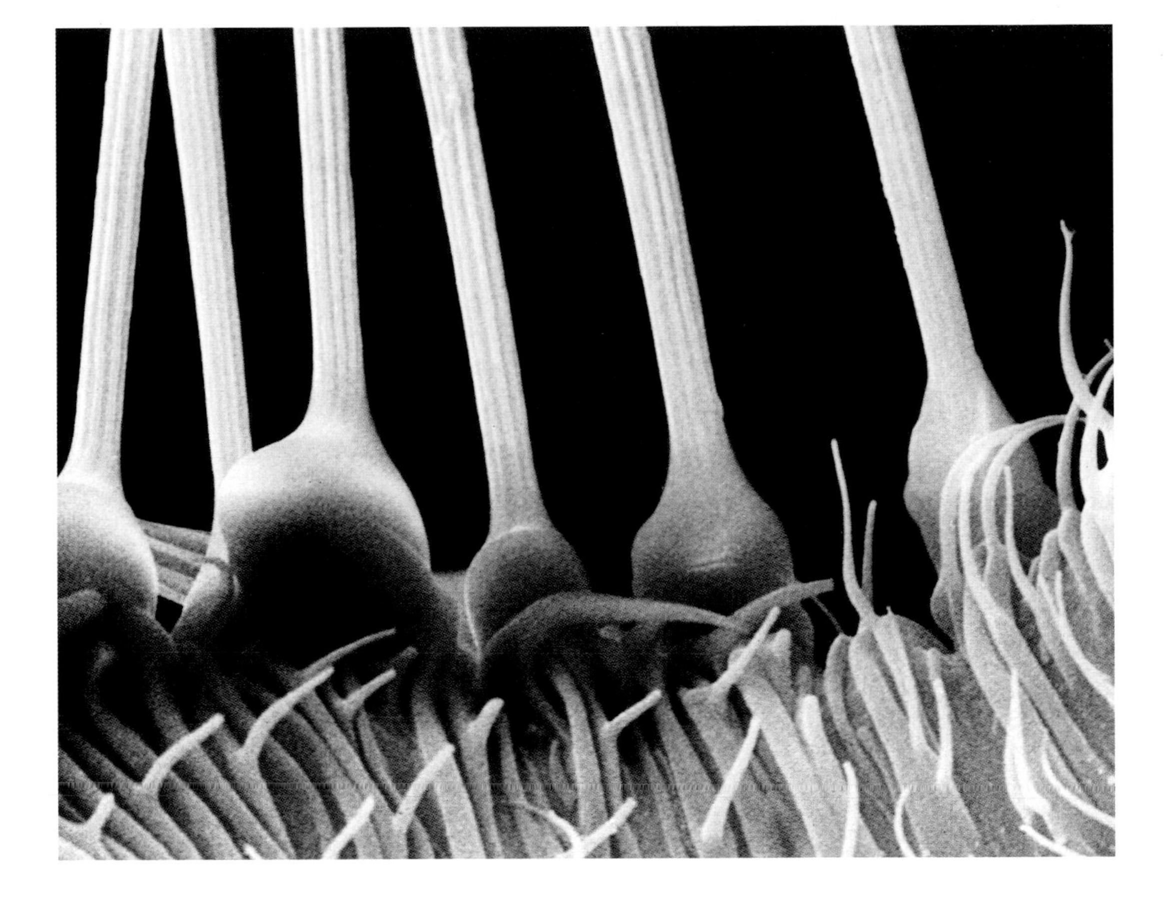

3.11 MOSQUITO WING SERIES: at the edge, x 3,000 (6)

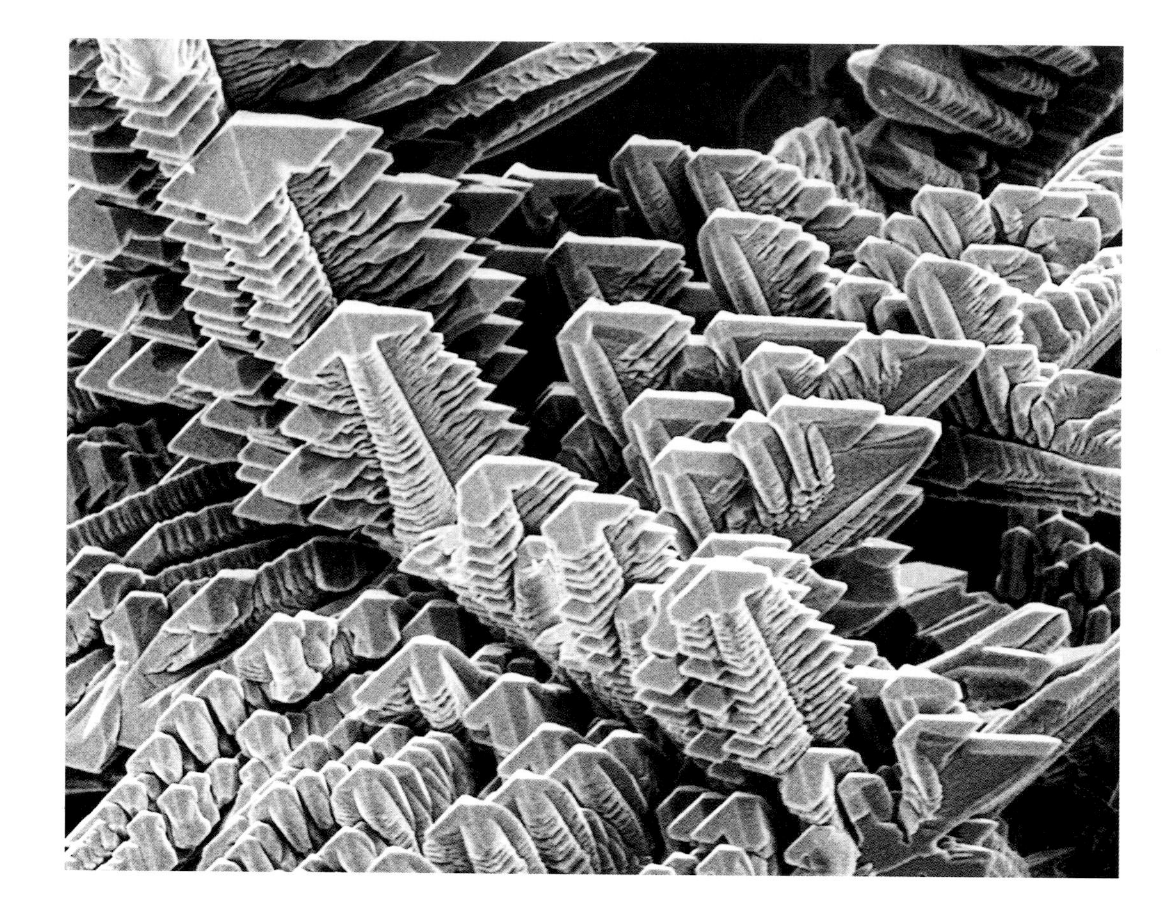

3.12 Surface detail on burned tungsten wire, x 700 (9)

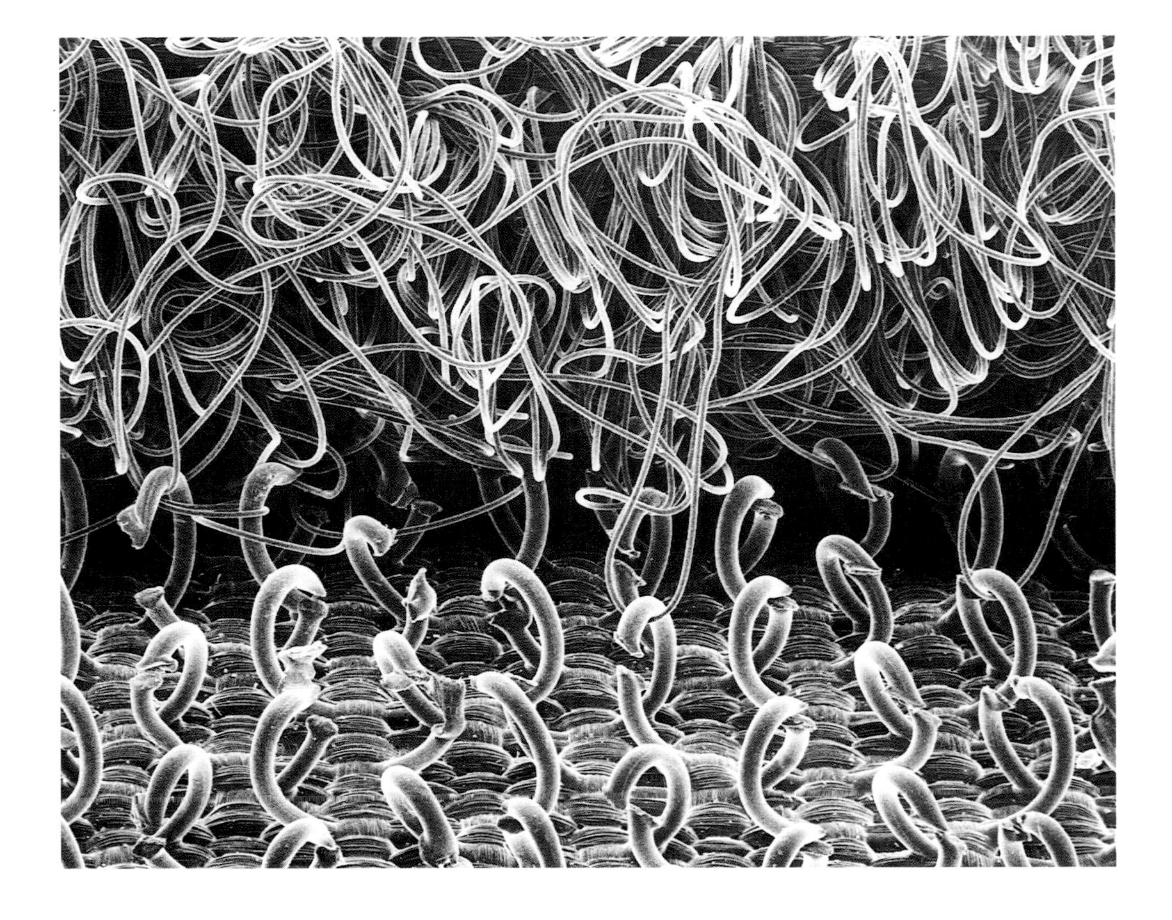

3.13 CAUGHT IN THE ACT: Velcro, joined, x 14 (24)

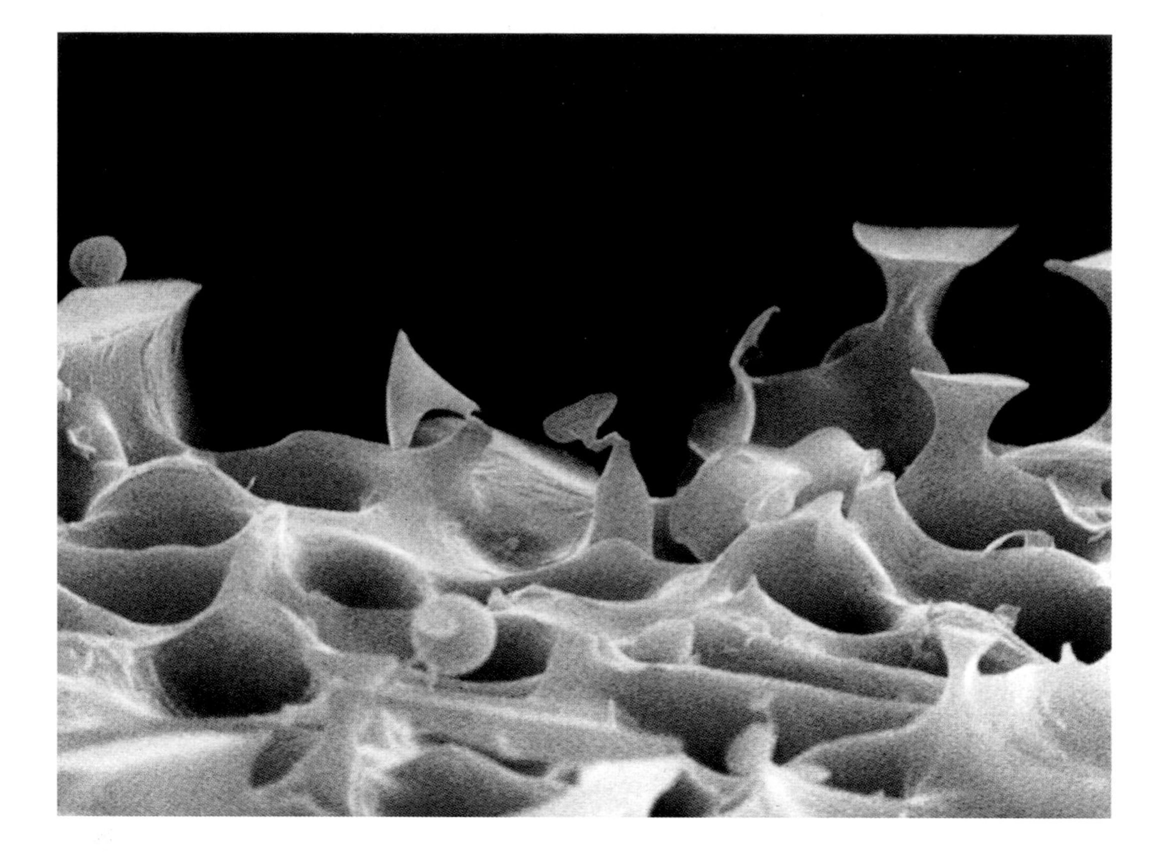

3.14 SALVADOR DALI REVISITED: the edge of a human protein, x 7,600 (25)

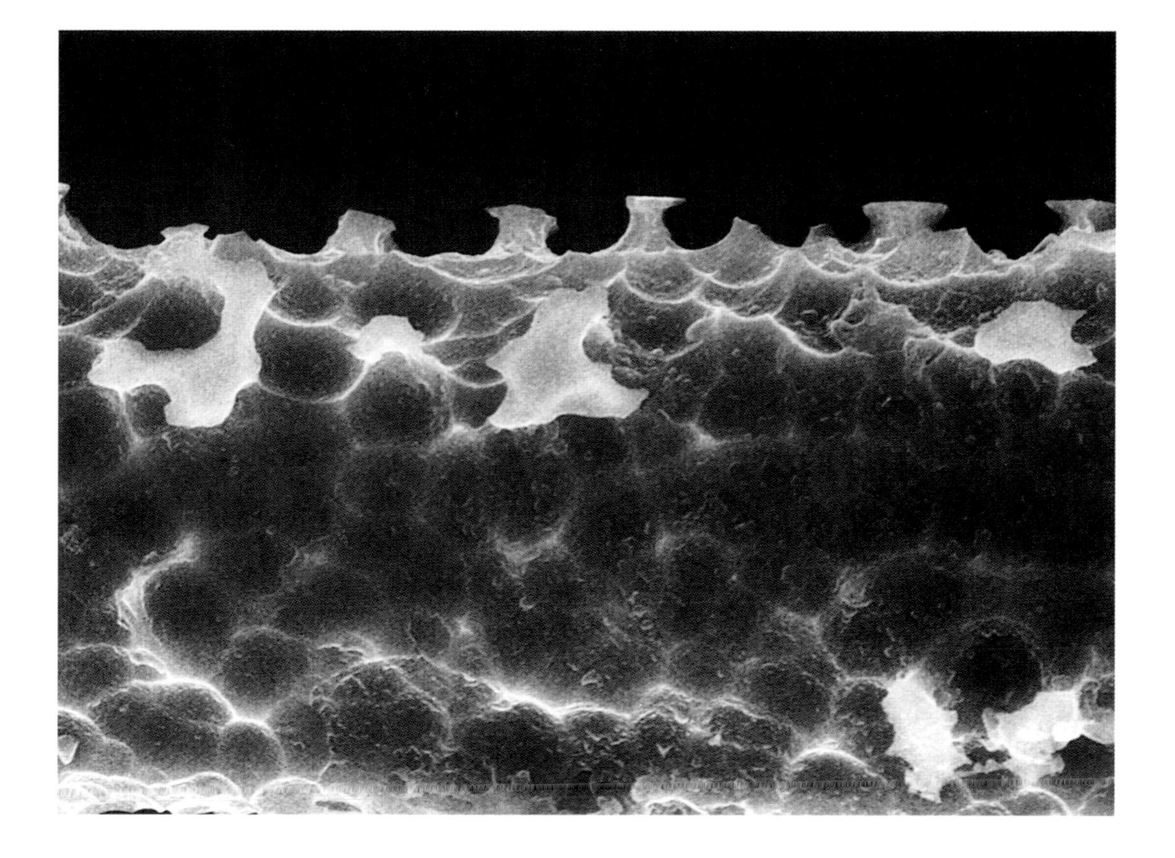

3.15 MICROTEKTITE: the edge of a drop of molten Earth-rock that fell to the sea and settled in sediment after a comet or asteroid crashed near Indonesia 800,000 years ago, x 850 (26)

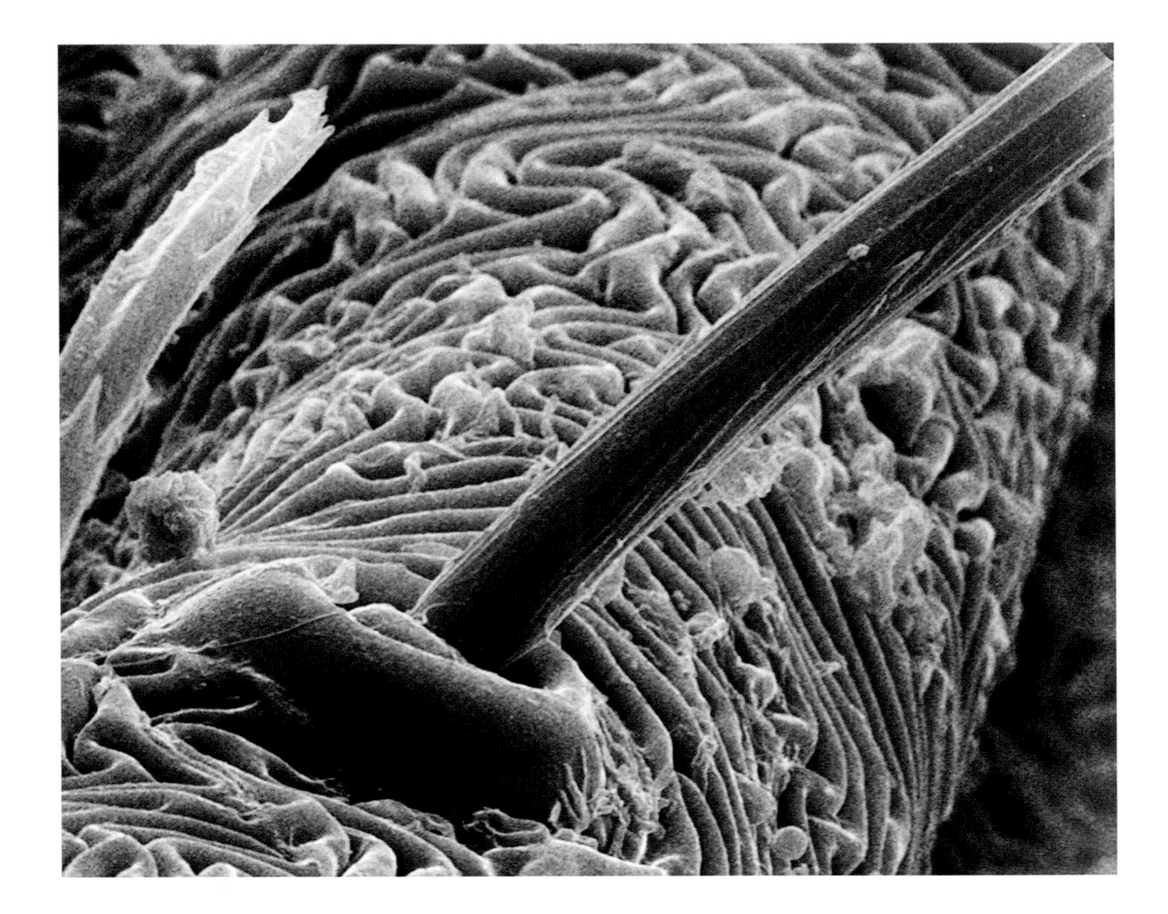

3.16 On the back of a bug, x 1,800

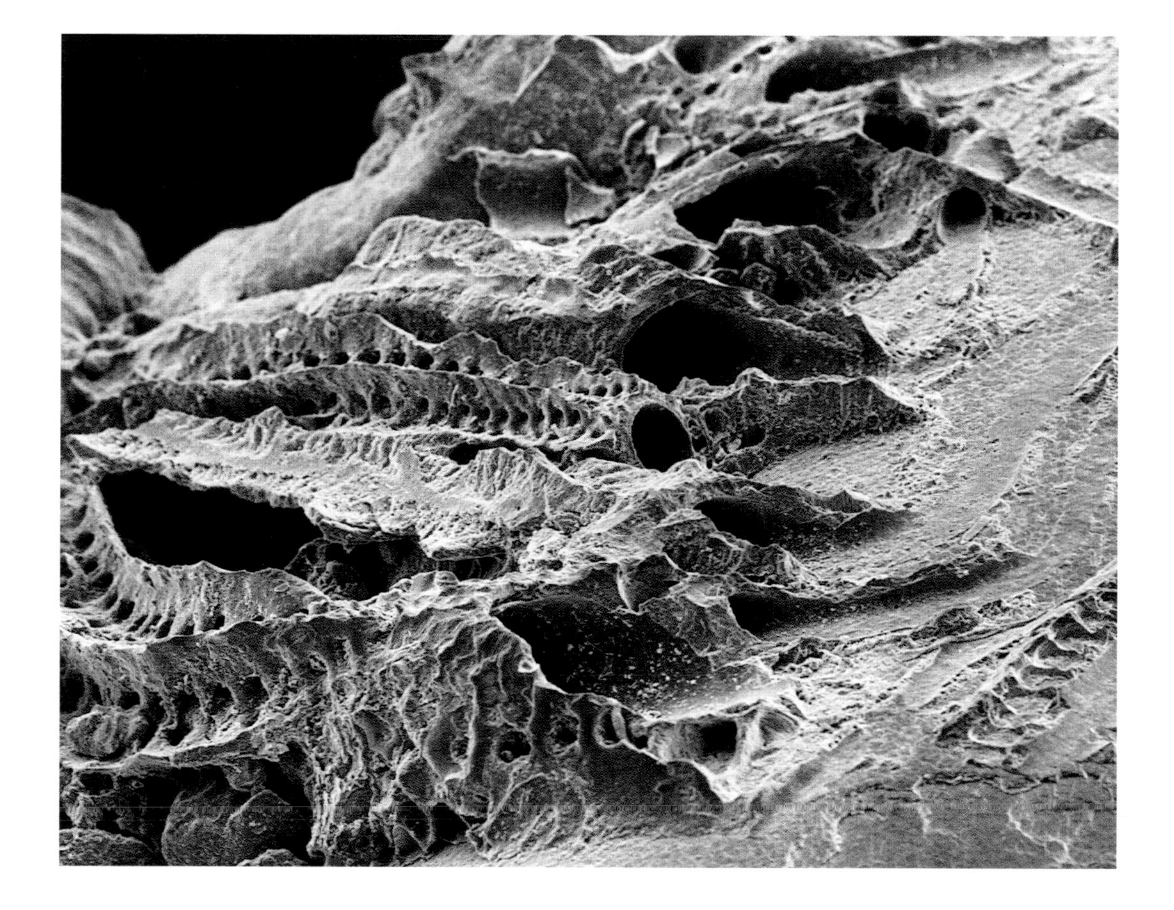

3.17 Tubes built by sea worms on a shell from Baja California, x 65 (27)

3.18 PETRIFIED LIGHTNING TRACK: grains of sand that fused around a tiny bolt of lightning, as it struck in the hills of North Carolina, x 25 (28)

3.19 MOSQUITO WING SERIES: at the edge, with a glow of electrons from over the horizon, x 7,000 (6)

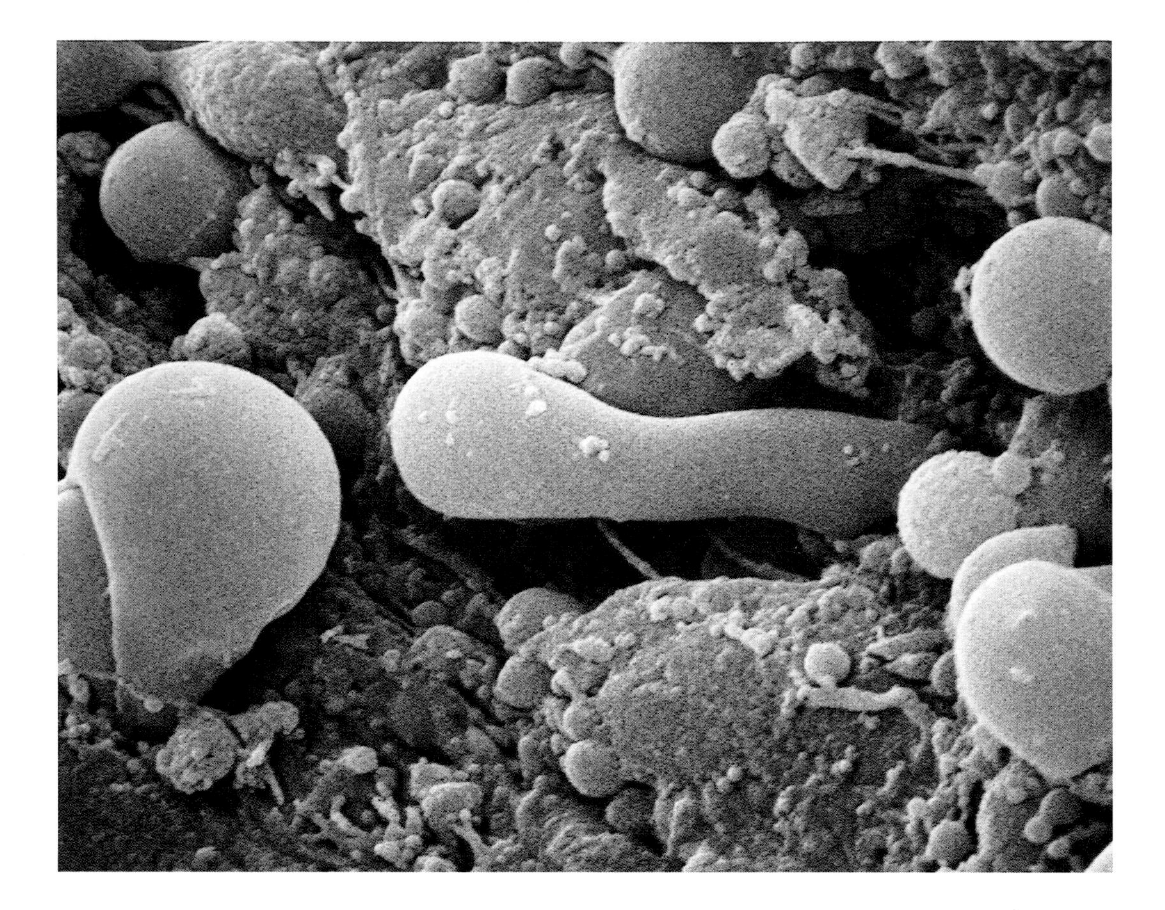

3.20 CAUGHT IN THE ACT: hormone excretion in a human thyroid gland, x 9,600 (29)

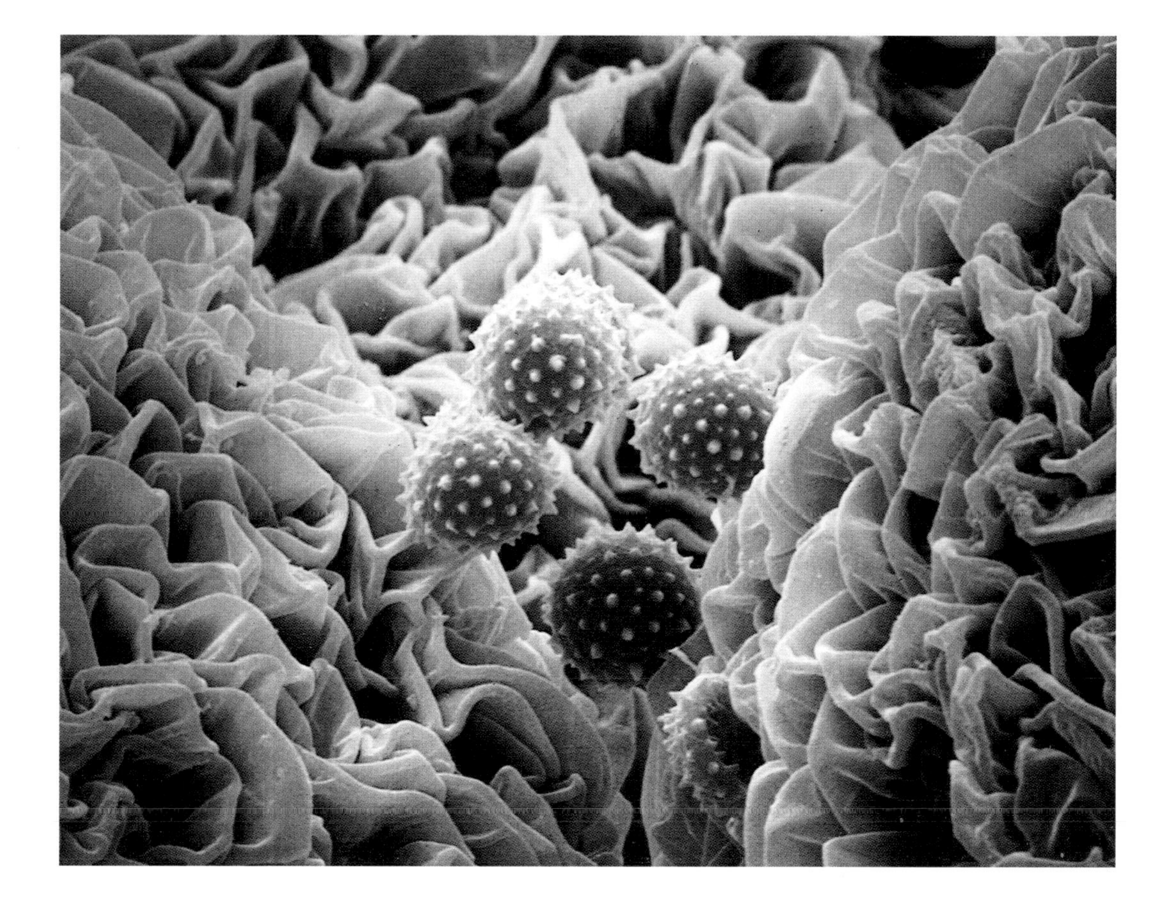

3.21 Pollen, suspended on a Jack-in-the-Pulpit, x 1,000 (23)

3.22 Mysterious formations on the surface of a sand grain taken from sediments beneath the Ross Sea, x 2,800 (30)

3.23 Scales on a butterfly wing, and the light of electrons through thin membrane, x 500 (6)

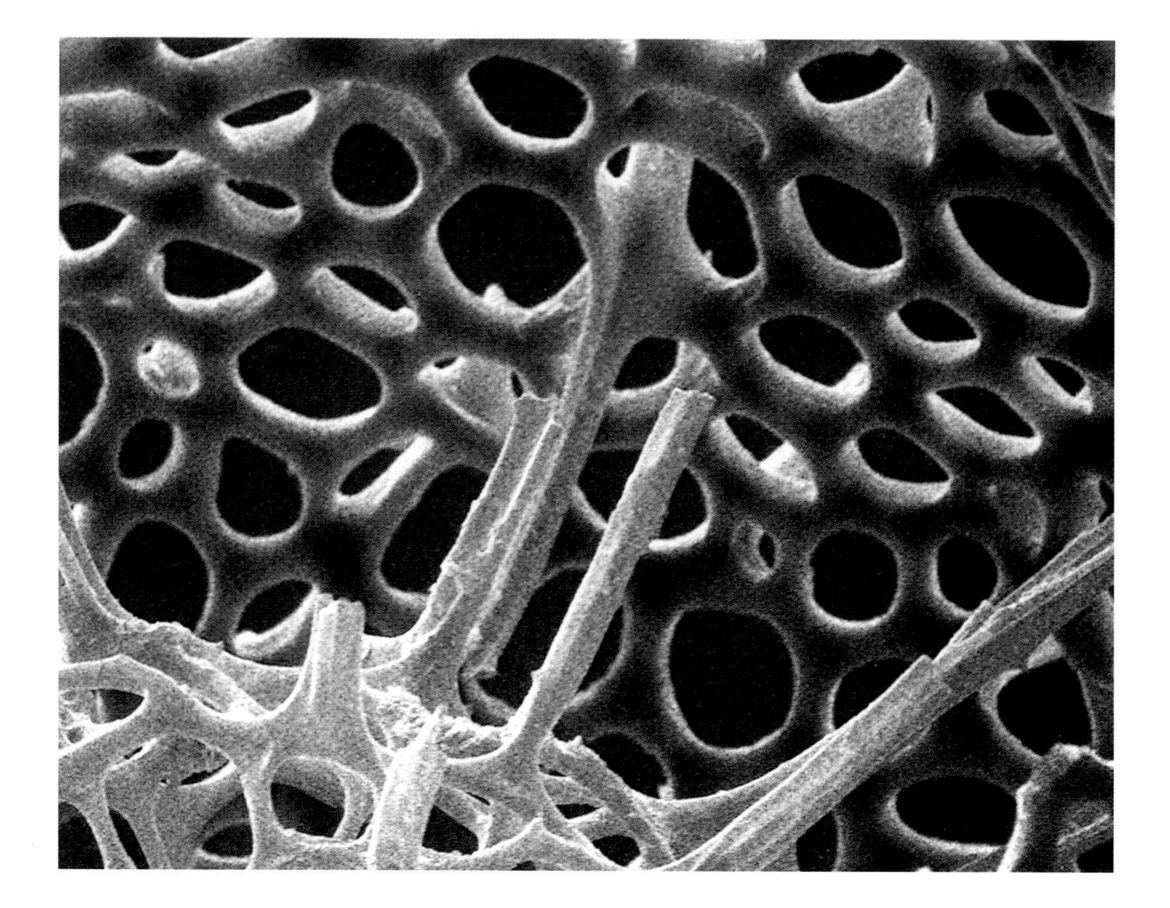

3.24 Looking at the world through a radiolarian, x 1,400 (1, 31)

3.25 Down feather, x 500

3.26 Betsy Beetle, Scene 3, x 800 (6)

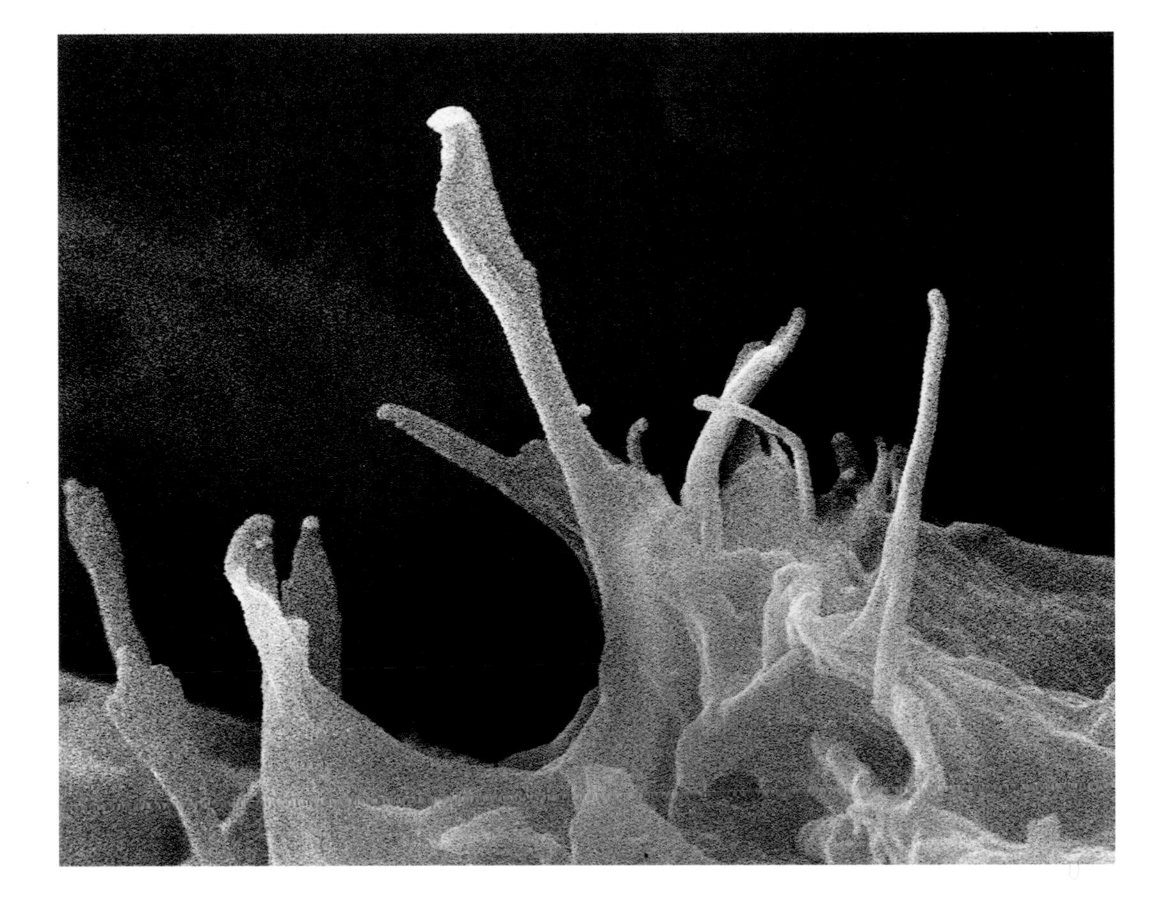

3.27 Mineral filaments reaching into a tiny crack in a fragment of rock dredged from a fracture in the floor of the Indian Ocean, x 20,000 (18)

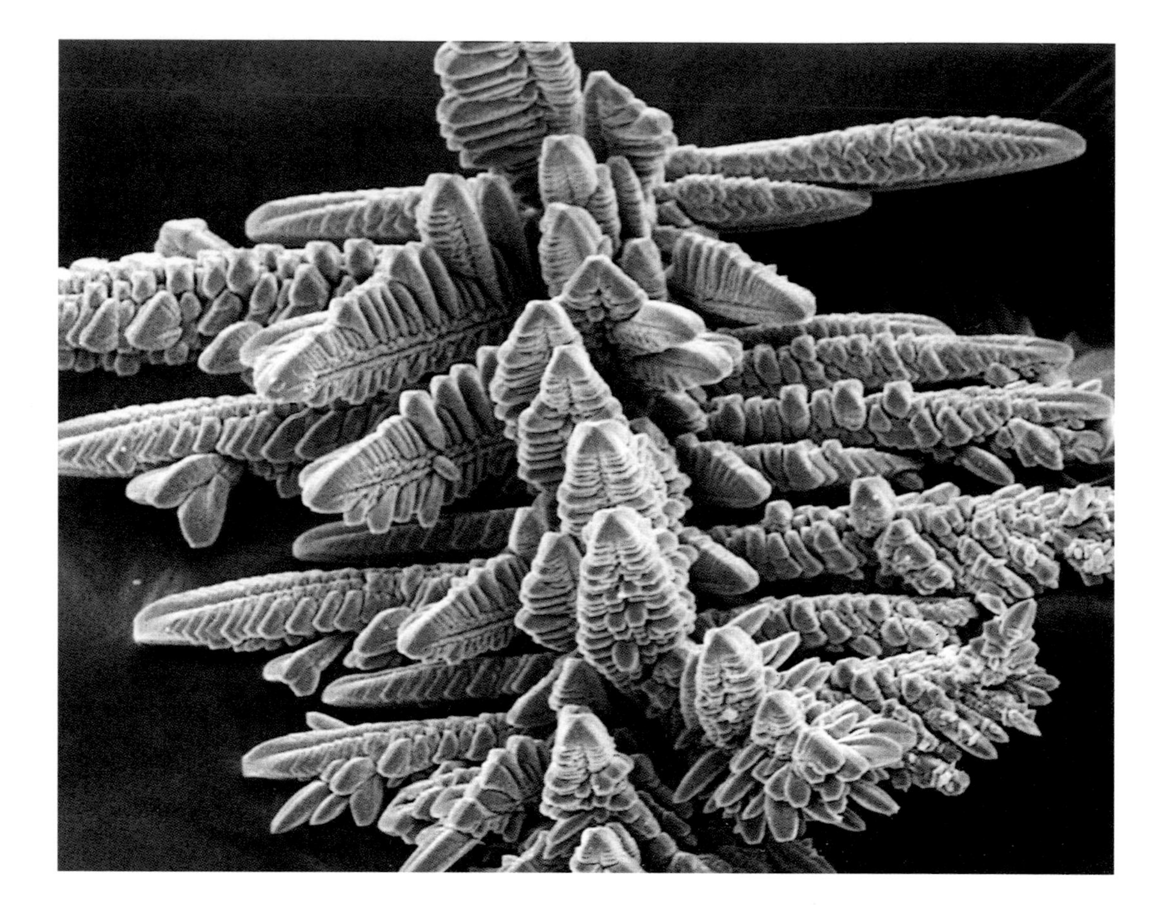

3.28 Residue of burned tungsten wire, x 600 (9)

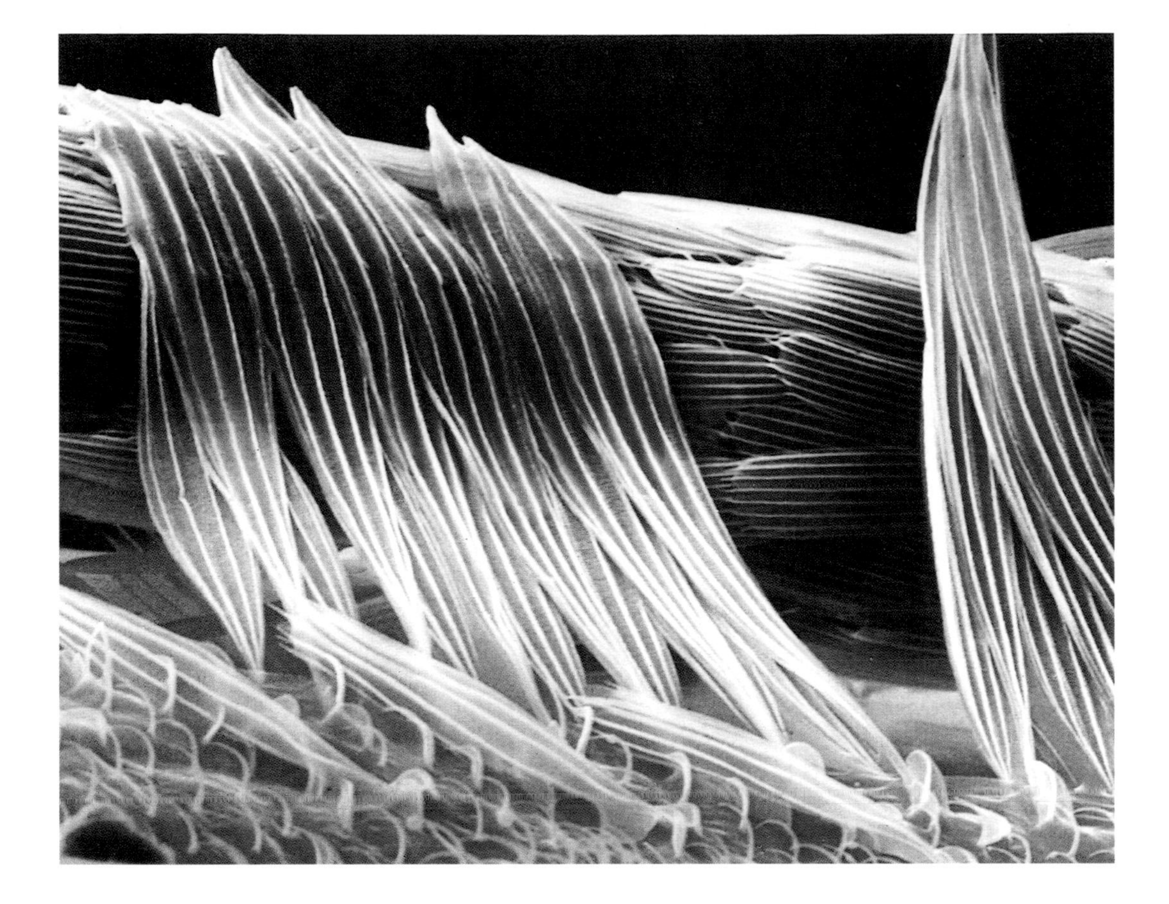

3.29 MOSQUITO WING SERIES: at the edge, x 1,100 (6)

KALEIDOSCOPE

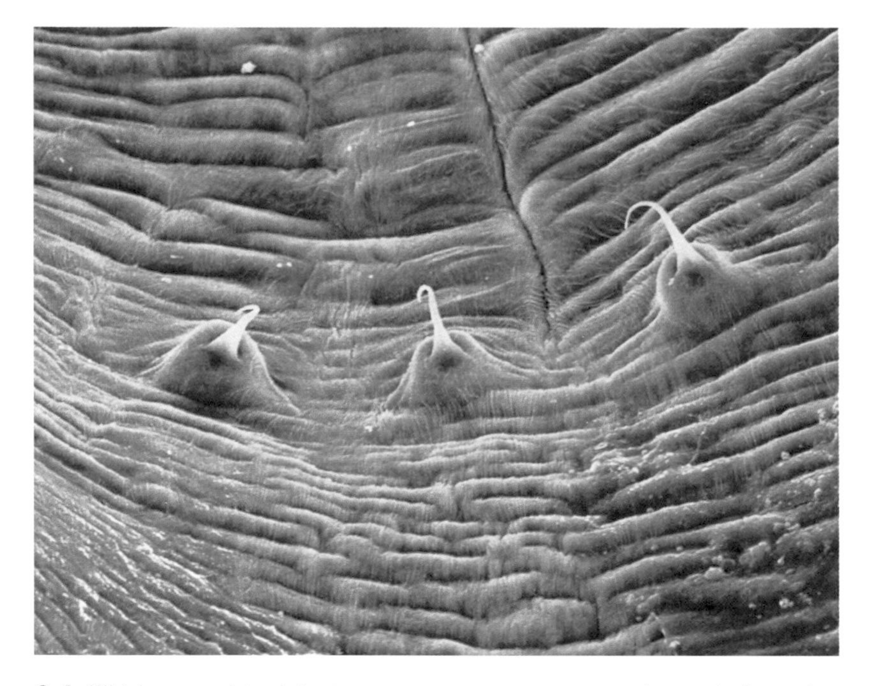

C.1 Whiskers used to detect ocean currents, on a young barnacle from the South Pacific, x 500 (32)

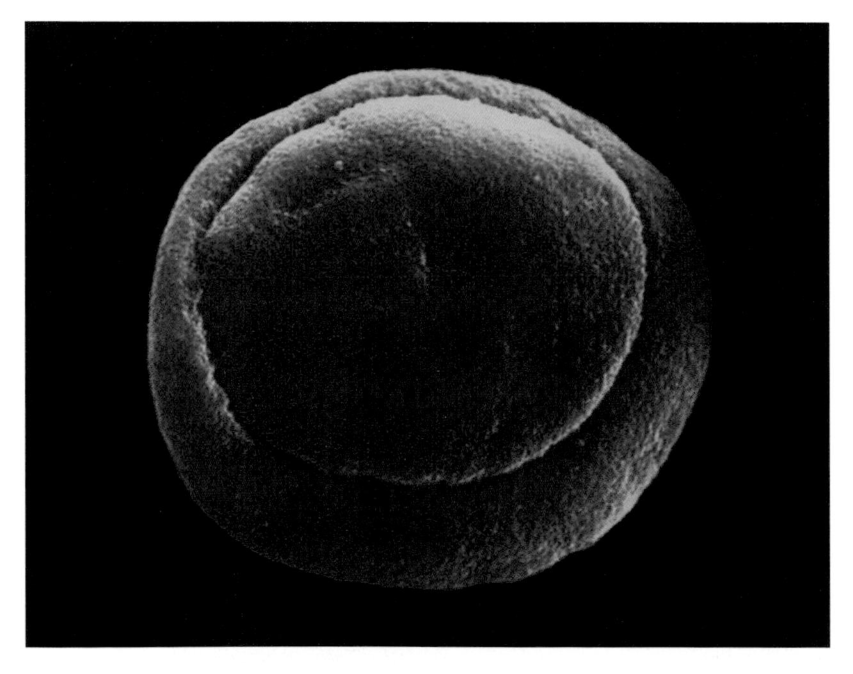

C.2 Fossil pollen grain from a conifer that lived 200 million years ago at a time of mass extinction, x 3,000 (59)

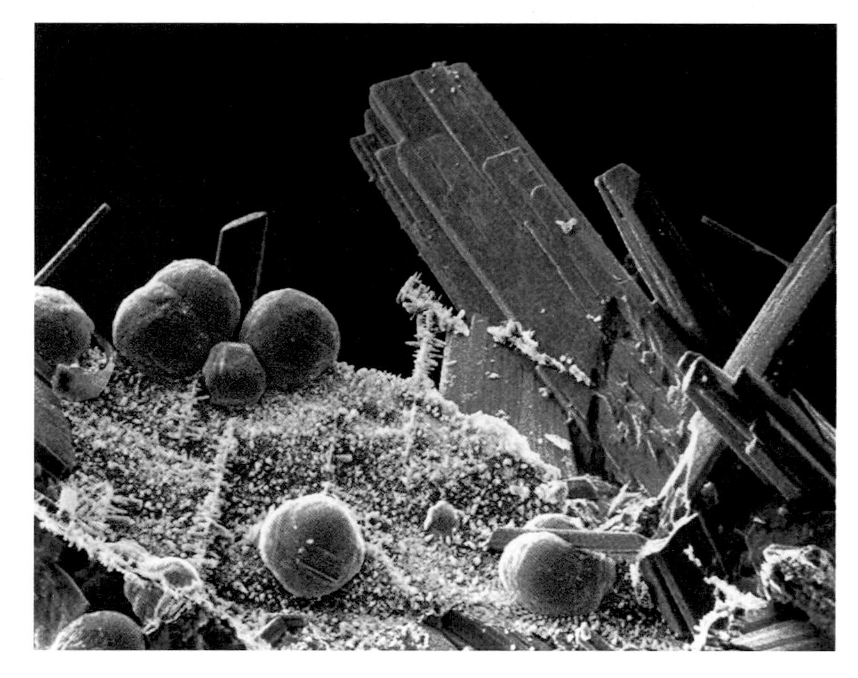

C.3 Franklin micromineral, x 200 (13)

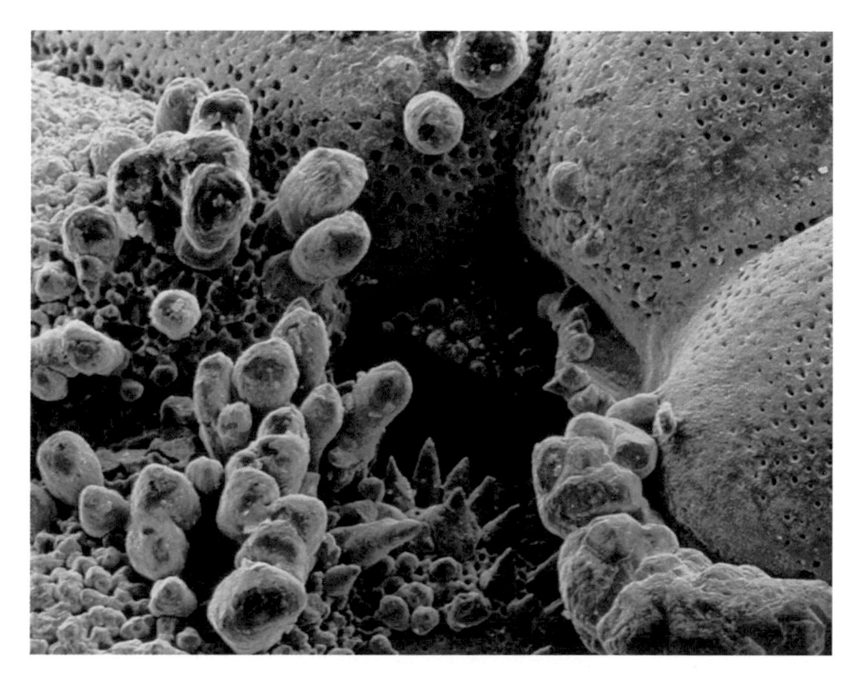

C.4 Center of the shell of a fossil foraminifer, x 200 (4, 56)

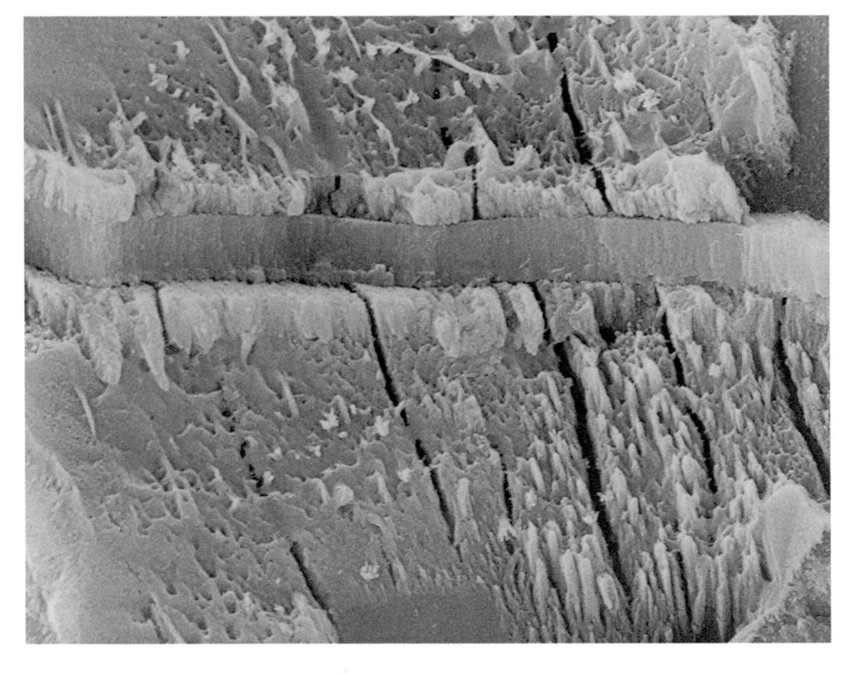

C.5 Chrysotile, cutting through a fragment of serpentine that was formed in the Earth's upper mantle and dredged from a deep trench in the floor of the Atlantic Ocean, x 3,800 (18)

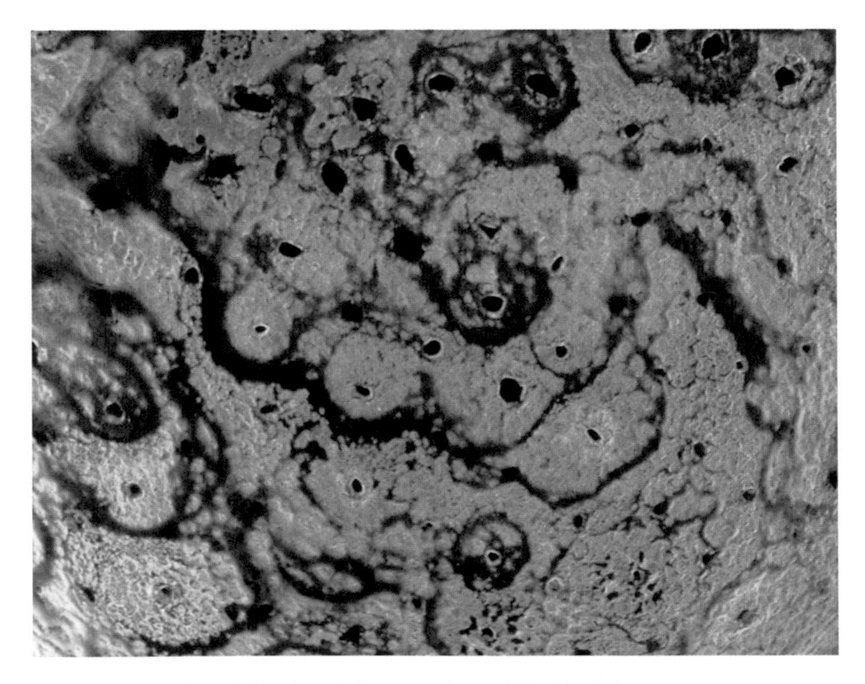

C.6 Surface detail of a fossil foraminifer, x 300 (4, 56)

C.7 Somewhere on a mosquito, x 2,900 (6)

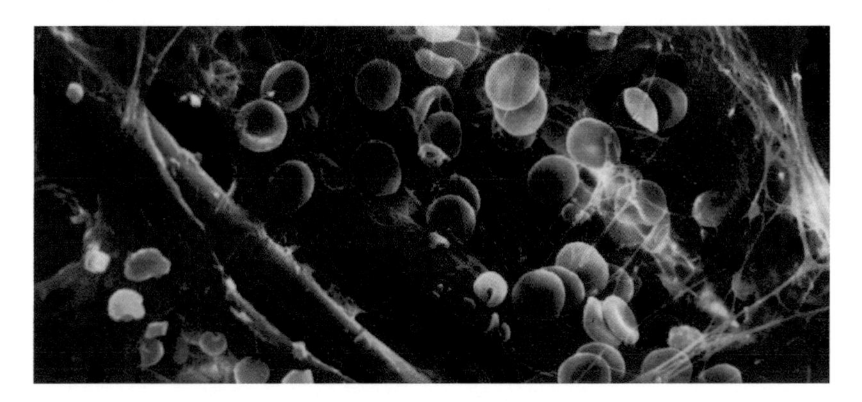

C.8 Arterial blood clot, x 1,400 (33)

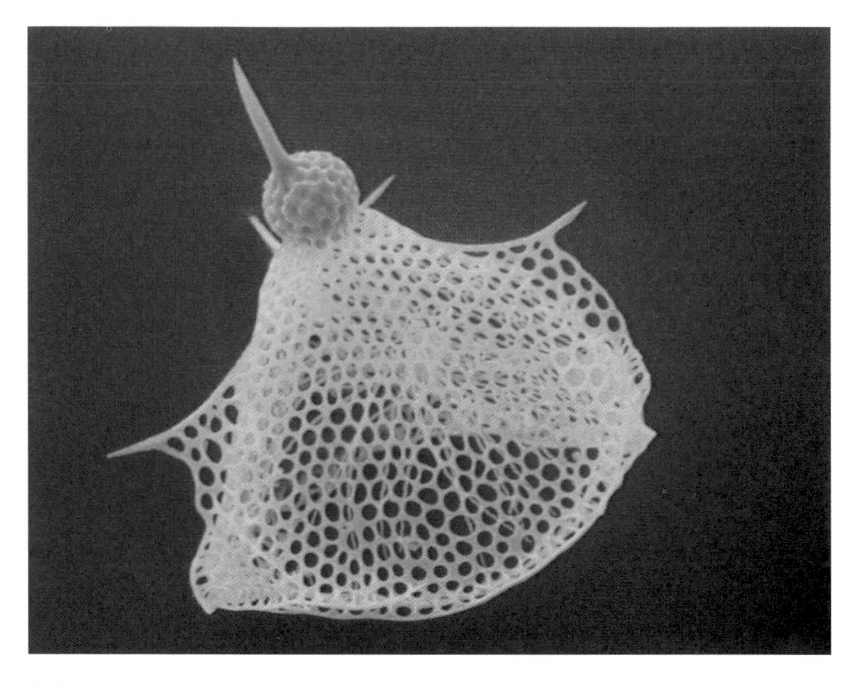

C.9 Caribbean radiolarian, x 450 (1, 51)

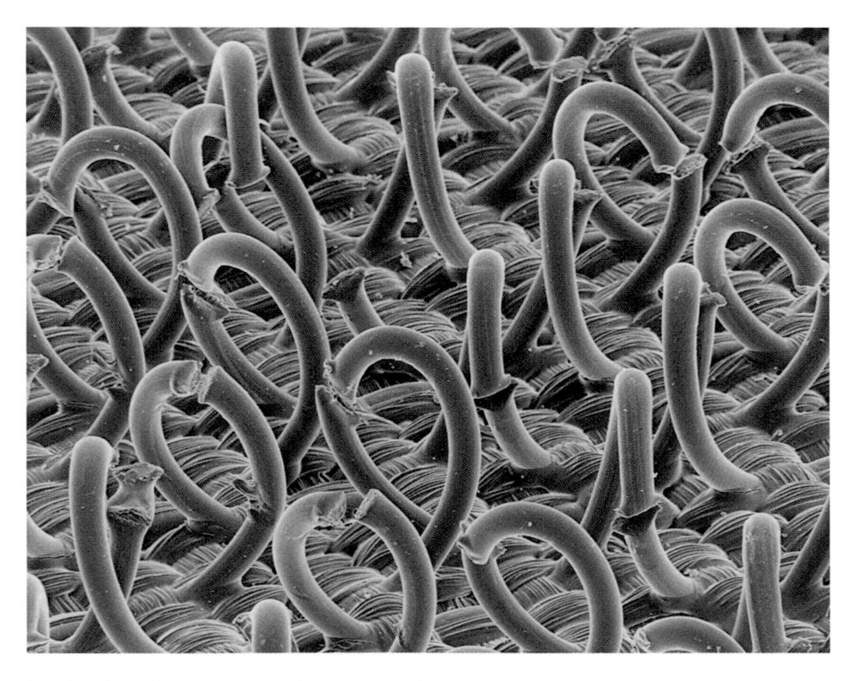

C.10 VELCRO: the hook side, x 25 (24)

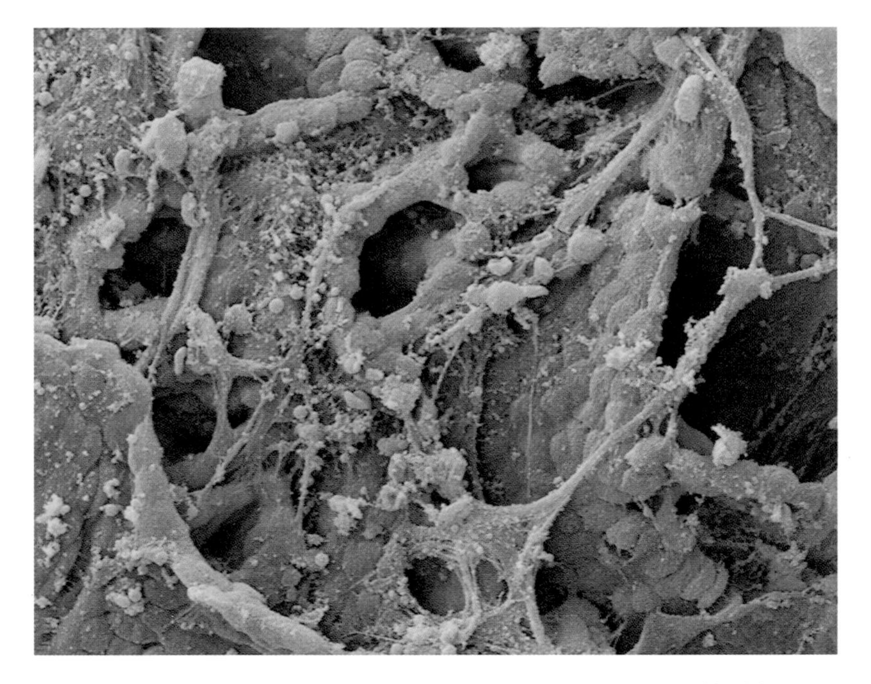

C.11 Secretion storage chambers in human thyroid tissue, x 600 (29)

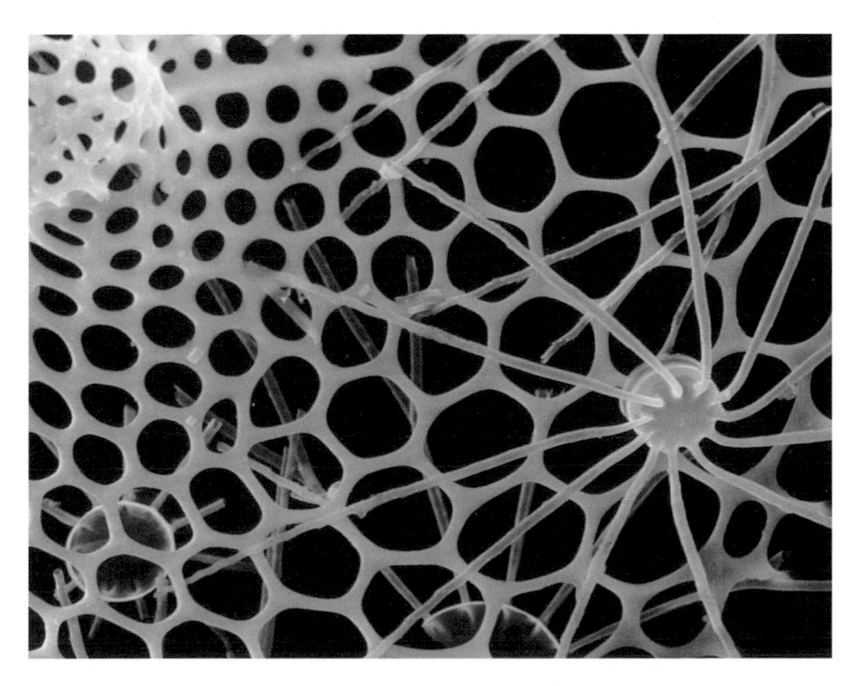

C.12 THREE DIATOMS ON A RADIOLARIAN: the radiolarian may have eaten the living cells of the diatoms and incorporated their shells into its own, x 1,400 (1, 51)

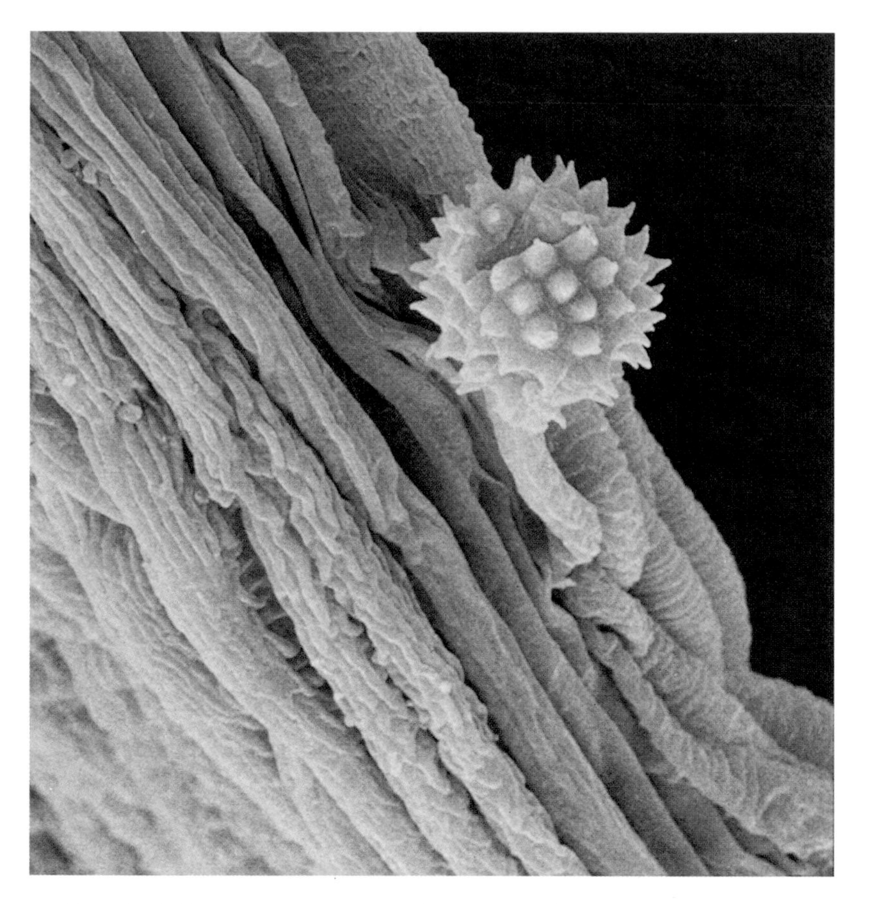

C.13 CAUGHT IN THE ACT: goldenrod pollen fertilizing a plant through the tube it grew after landing, x 1,600 (47)

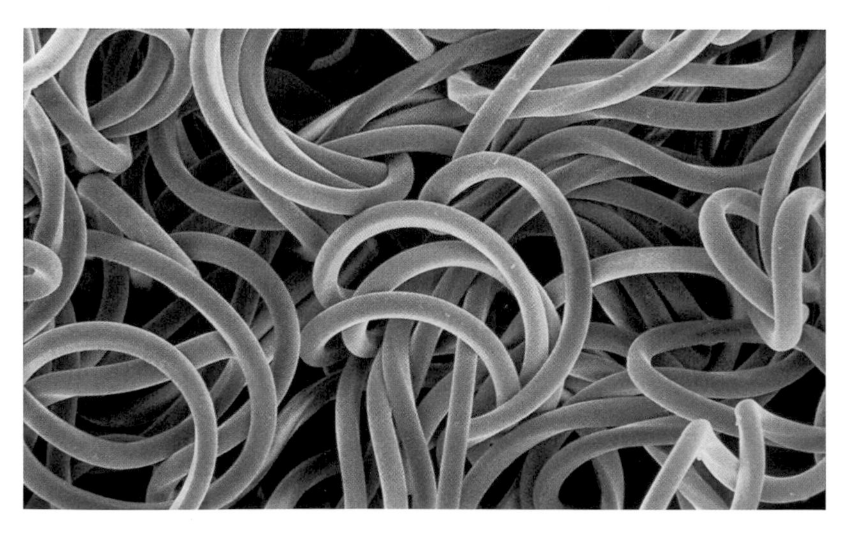

C.14 Nylon stocking, x 170 (60)

C.15 MOSQUITO WING SERIES: surface detail, x 1,600 (6)

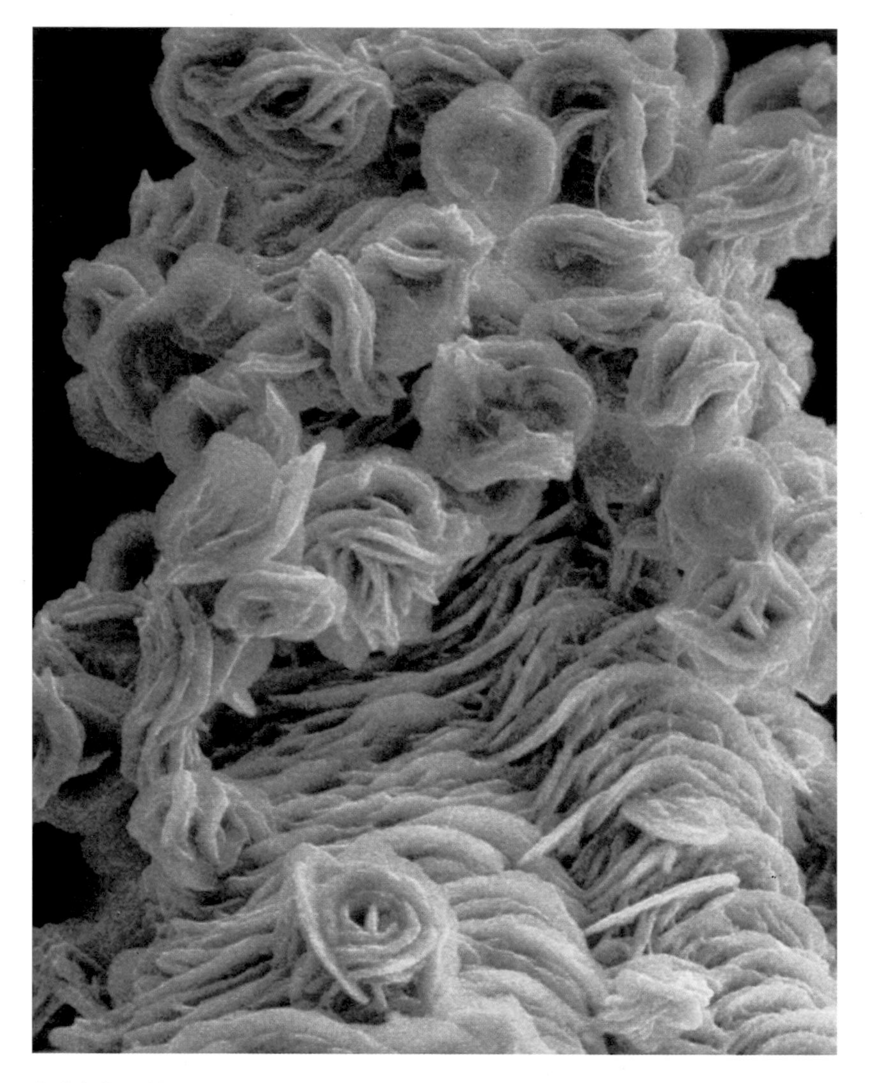

C.16 Franklin micromineral (a bug's eye view), x 1,200 (13)

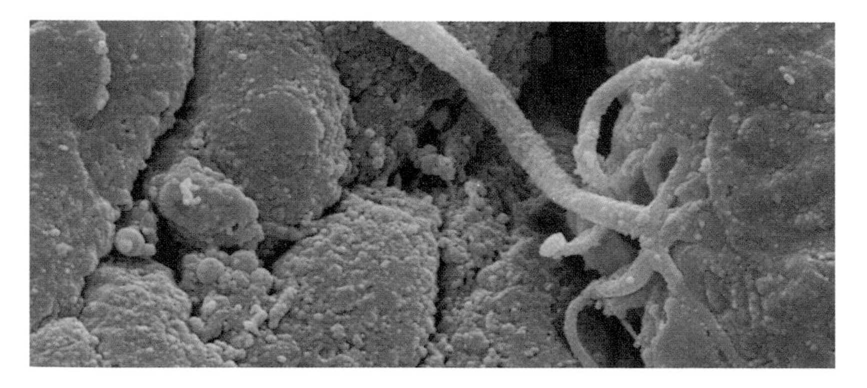

C.17 Blood vessel anchored in a network of capillaries running through thyroid tissue, x 4,800 (29)

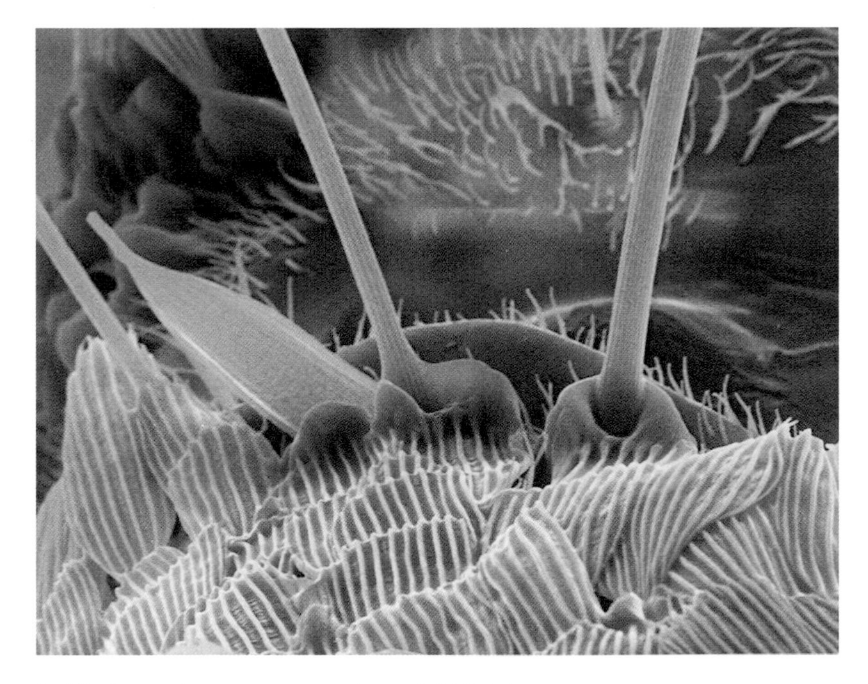

C.18 Somewhere on a mosquito, x 1,100 (6)

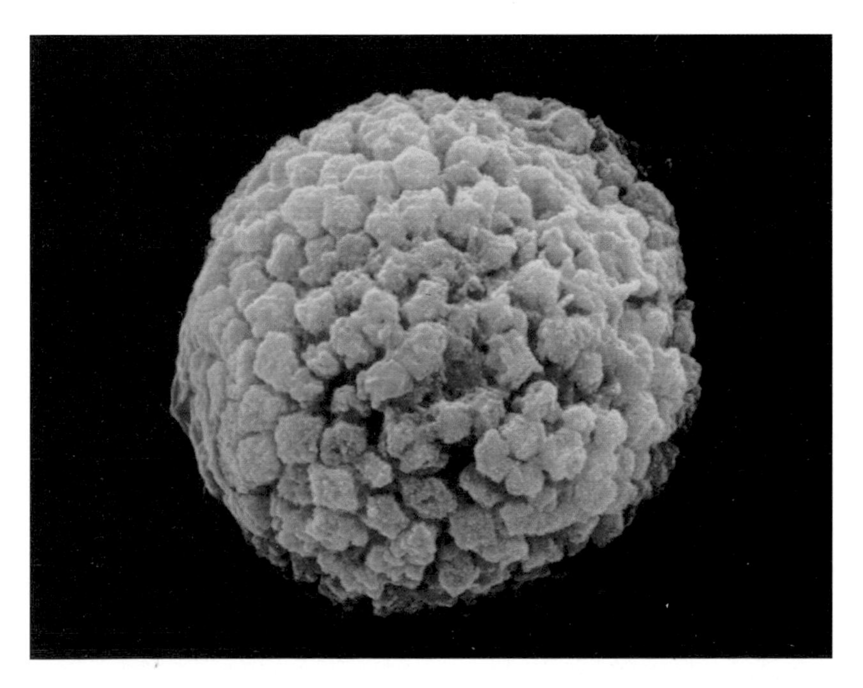

C.19 Pyrite dredged from a trough in the sea floor beneath the equatorial Atlantic, x 5,300 (18)

C.20 Coccolithophorid, whimsically colored (with apologies to Mother Nature), x 10,500 (3, 41)

C.21 Imprint of the base of a young barnacle where it was once attached to a larger barnacle in the sea near Tahiti, x 30 (32)

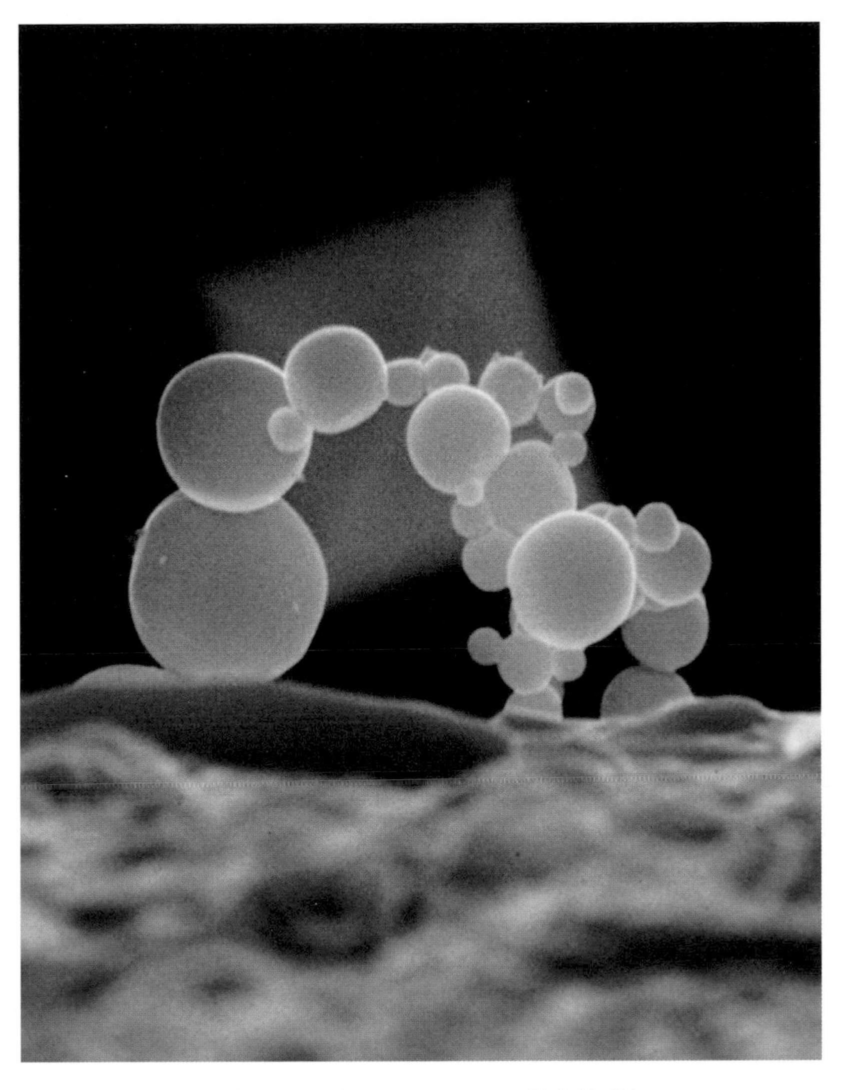

C.22 At the edge of a burned tungsten wire, x 24,000 (9)

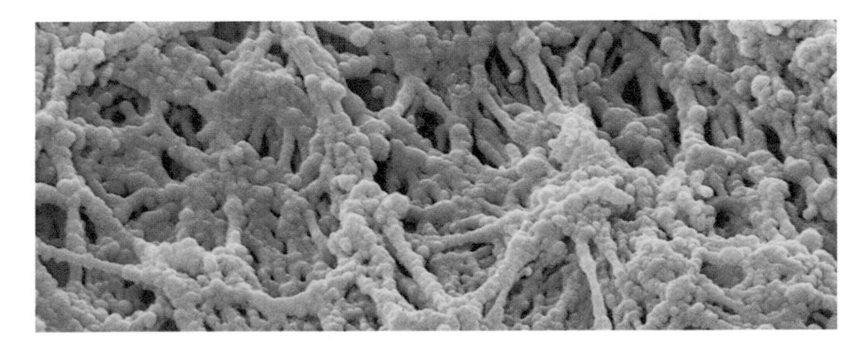

C.23 SUPPORT AND PROTECTION: connective tissue from a human thyroid, x 11,500 (29)

C.24 LABORATORY CRYSTALLIZATION OF SILICATE MAGMA: the natural forms of these lab-made minerals may have come from an ocean of lava, deep in a once-hotter Earth, x 2,500 (44)

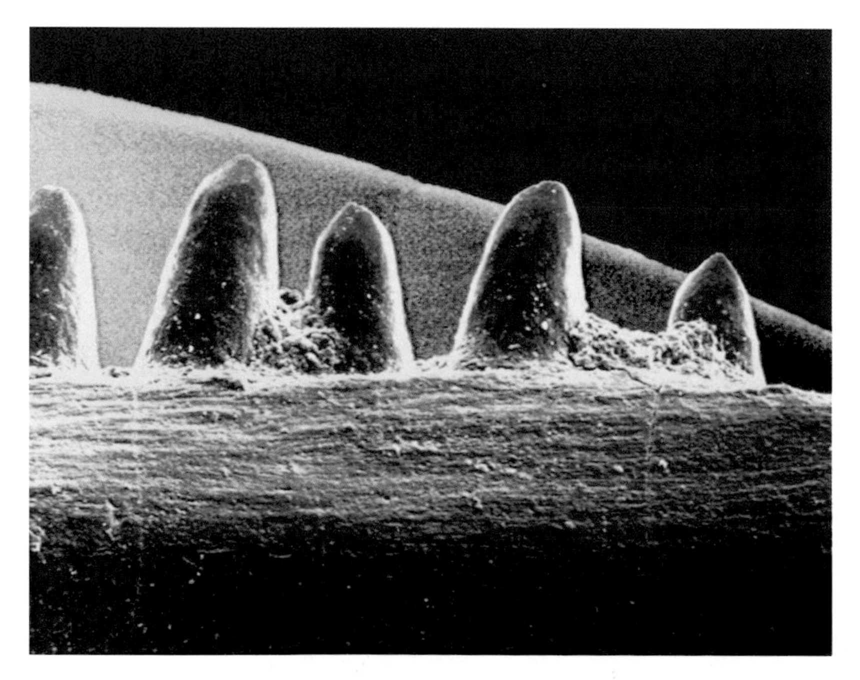

C.25 Teeth of a 225-million-year-old reptile that lived in the shadow of dinosaurs, x 125 (61)

C.26 Tropical coccoliths, x 14,500 (3, 17)

C.27 Two scales on a butterfly's wing, one peeling away, x 2,500 (6)

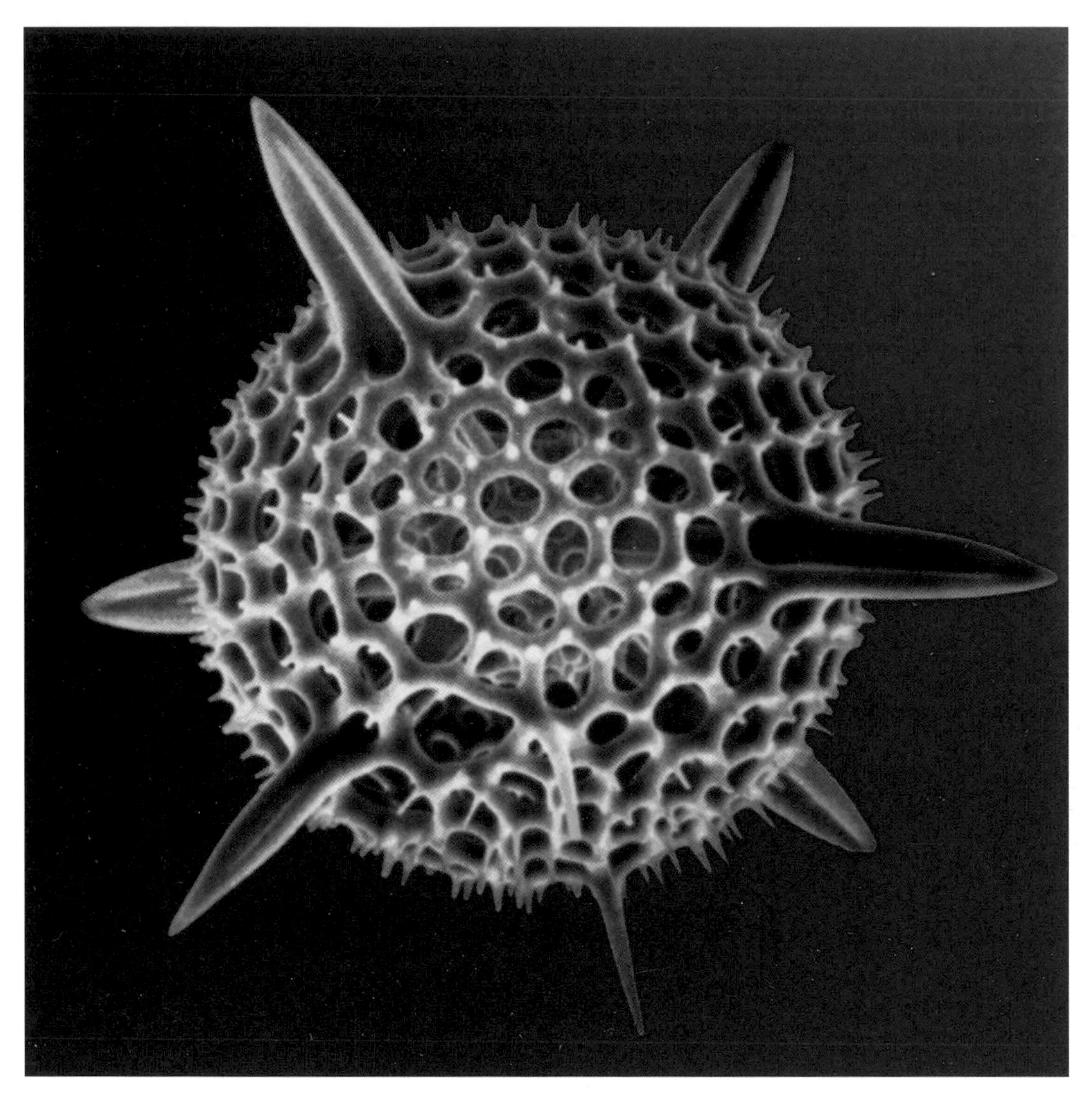

C.28 Fossil radiolarian, x 450 (1, 31)

4.1

4 · PATTERNS & TEXTURES

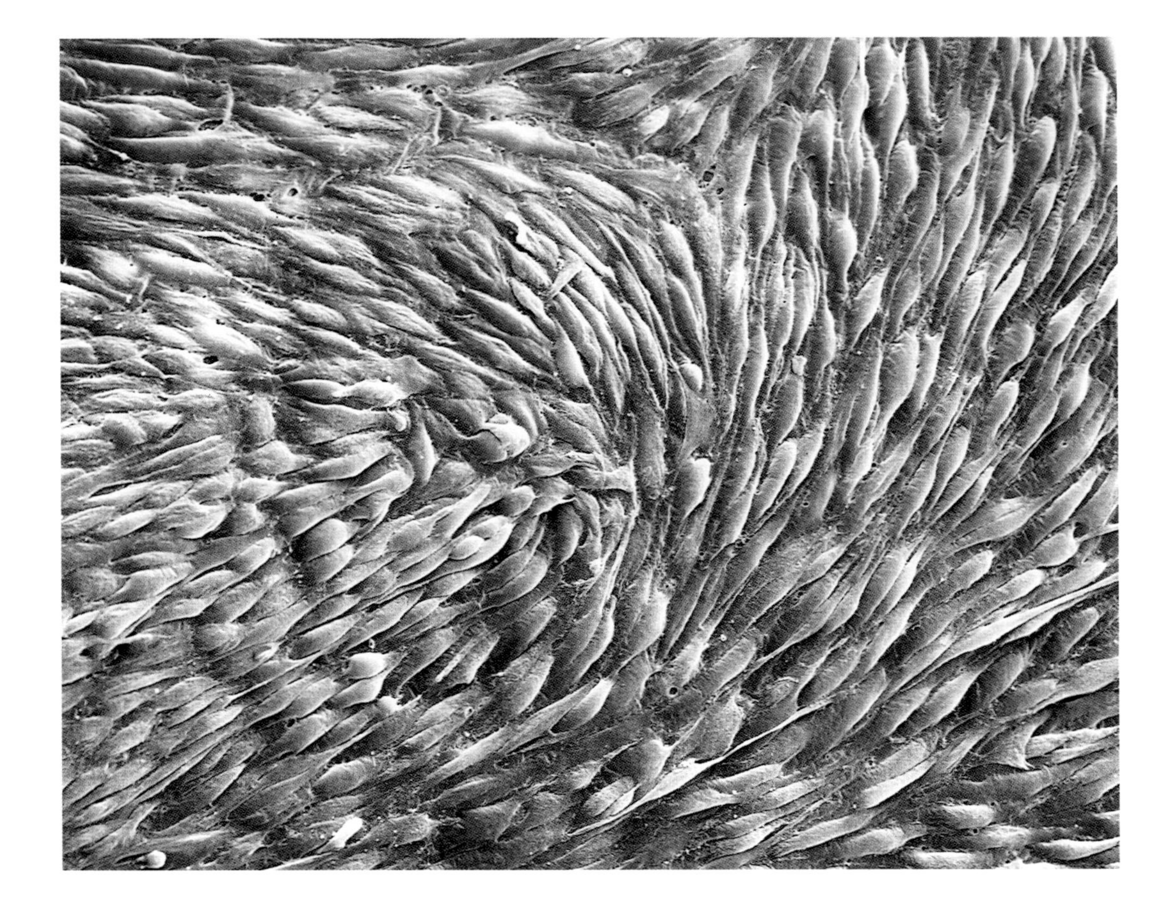

4.2 The cellular lining of an artery, x 400 (33)

4.3 Surface detail of copper, x 7,000 (34)

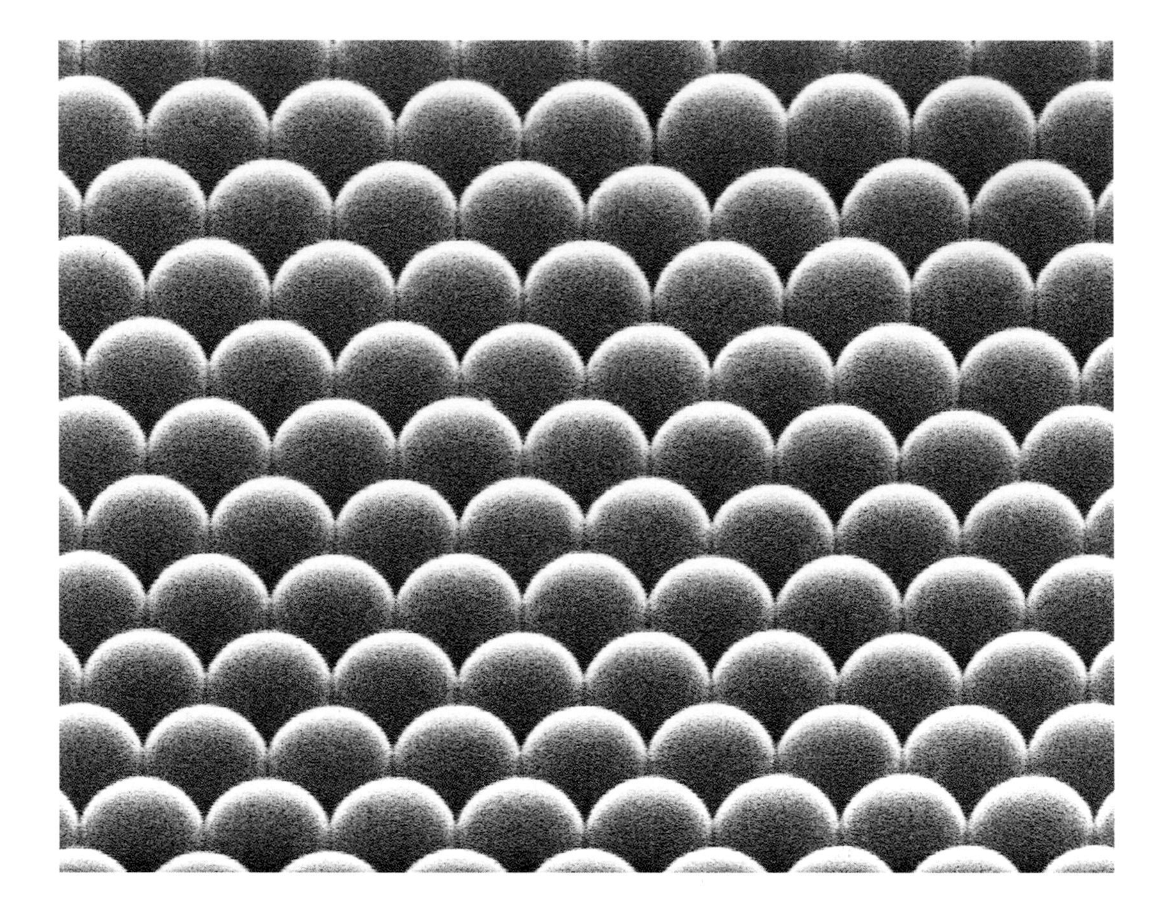

4.4 ENGINEERED UNIFORMITY: latex particles, x 48,000 (8)

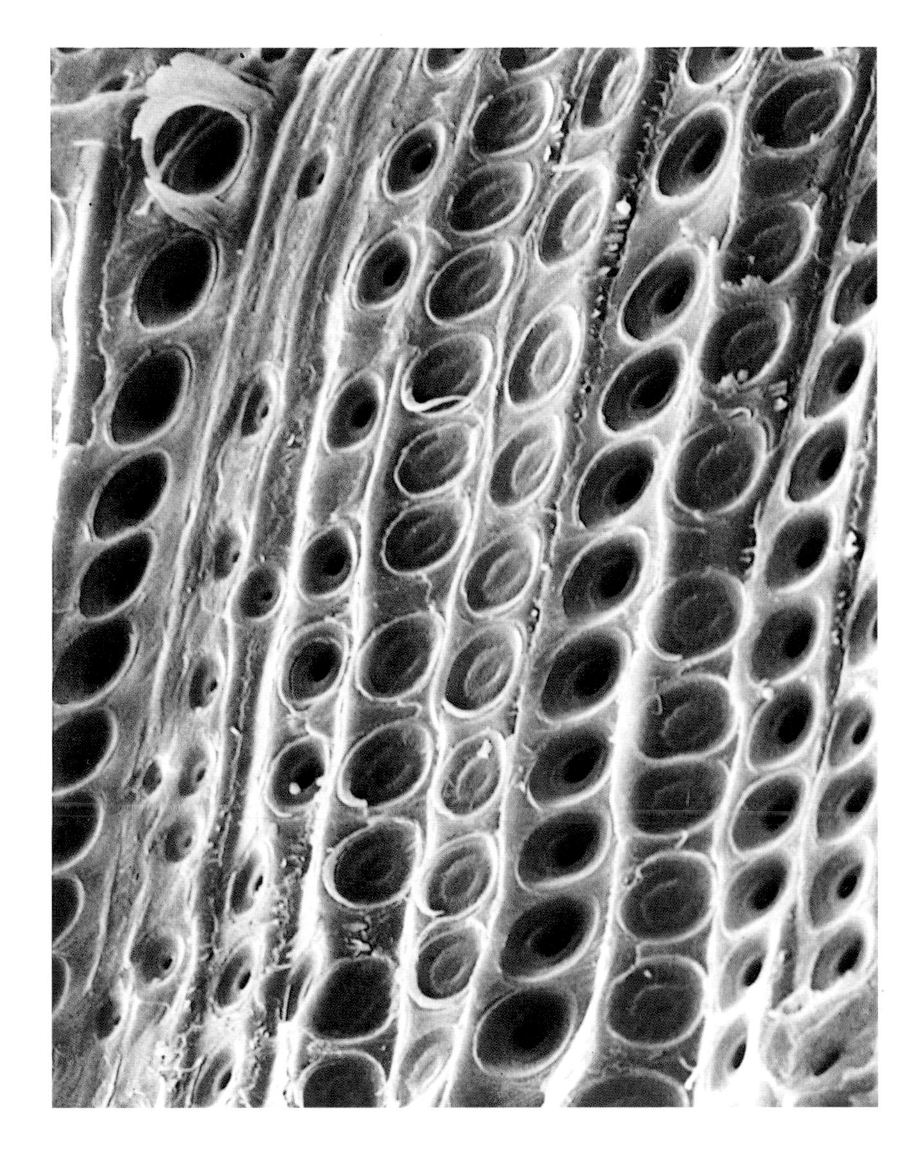

4.5 Pits in the cells of a Canadian conifer, x 900 (35)

4.6 Worm tube pattern 1, x 50 (27)

4.7 Worm tube pattern 2, x 35 (27)

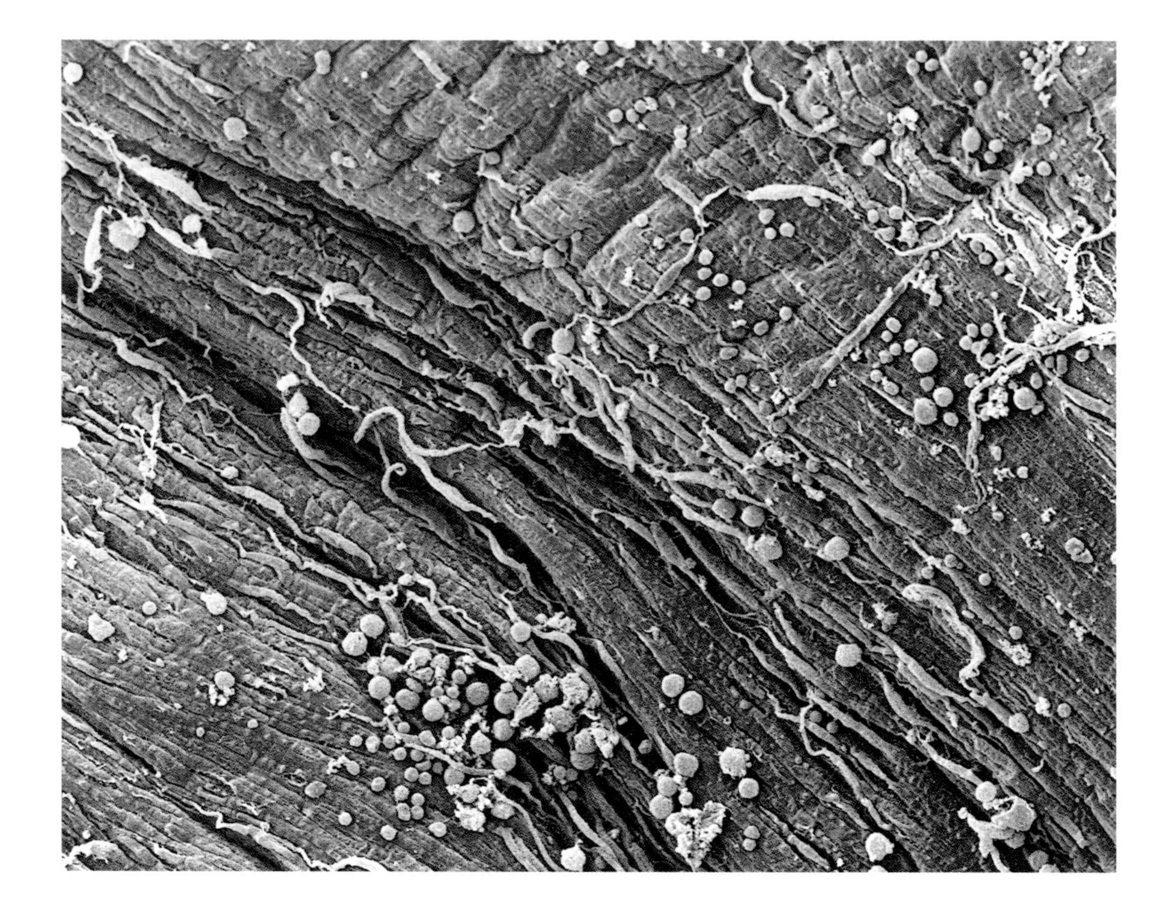

4.8 Inside an artery in a human lung, x 1,700 (20)

4.9 A close look at wood from a 2000-year-old Huon pine, which is still living in the Tasmanian rain forest, x 300 (36)

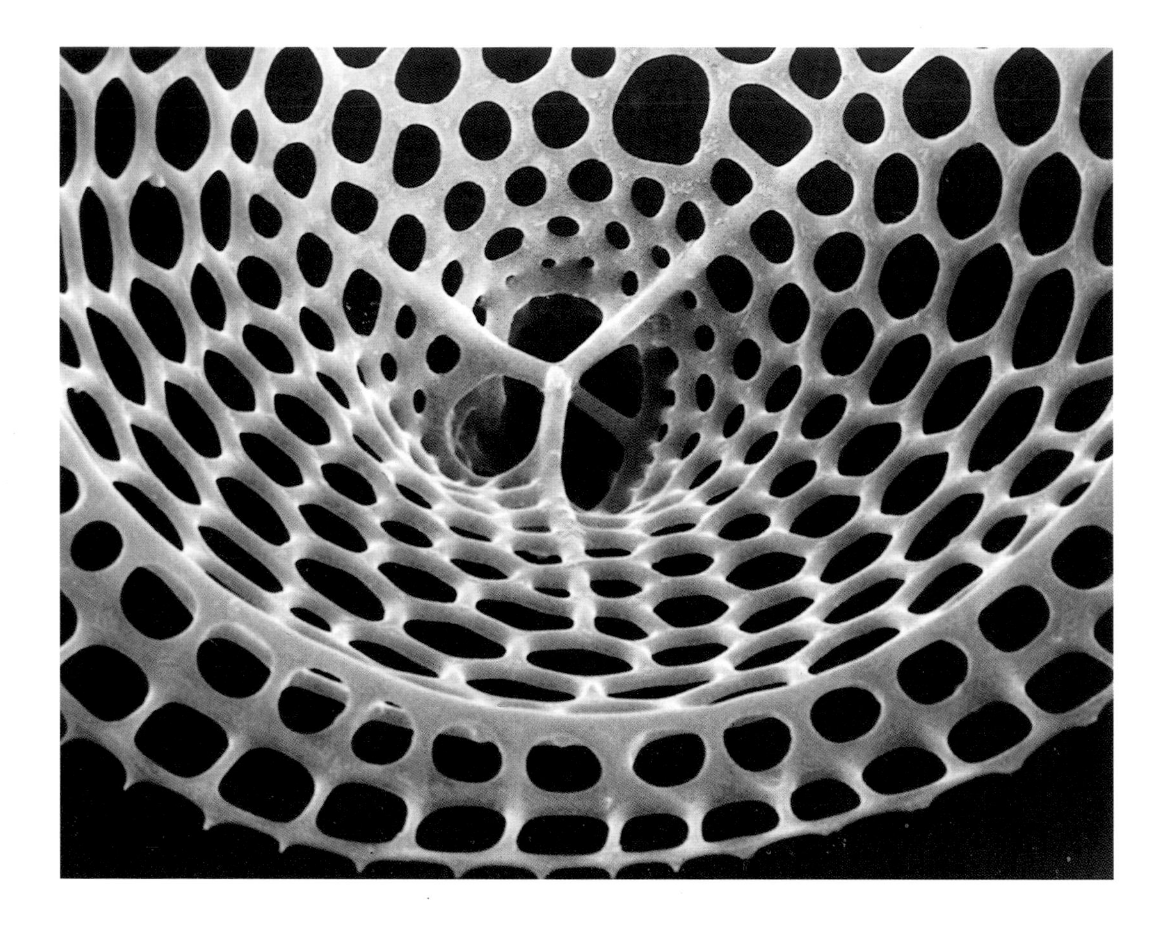

4.10 Radiolarian detail, x 1,200 (1, 31)

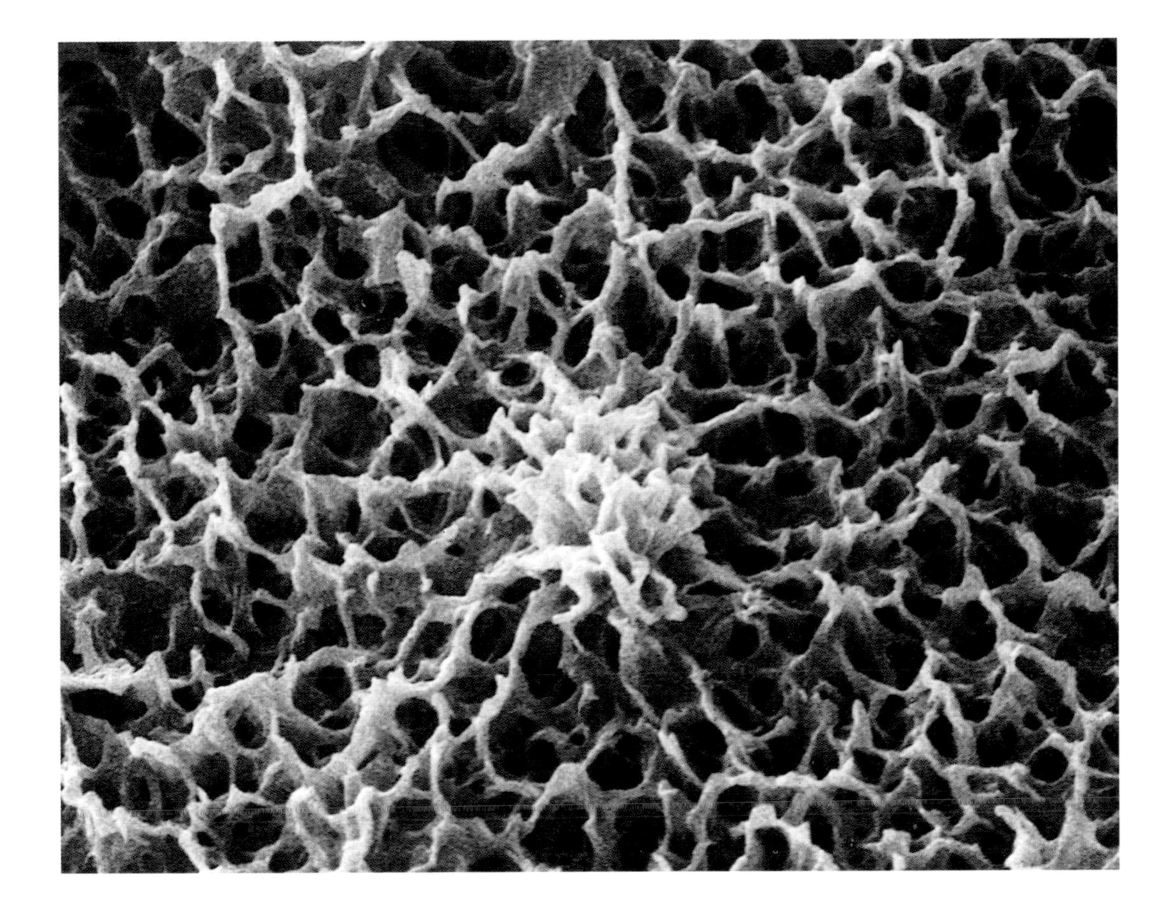

4.11 Serpentine, derived from the Earth's upper mantle and dredged from a fracture in the floor of the Indian Ocean, x 6,000 (18)

4.12 STOMATA ON A CACTUS: plants breathe through these openings, which conserve moisture by closing on hot or dry days, x 200

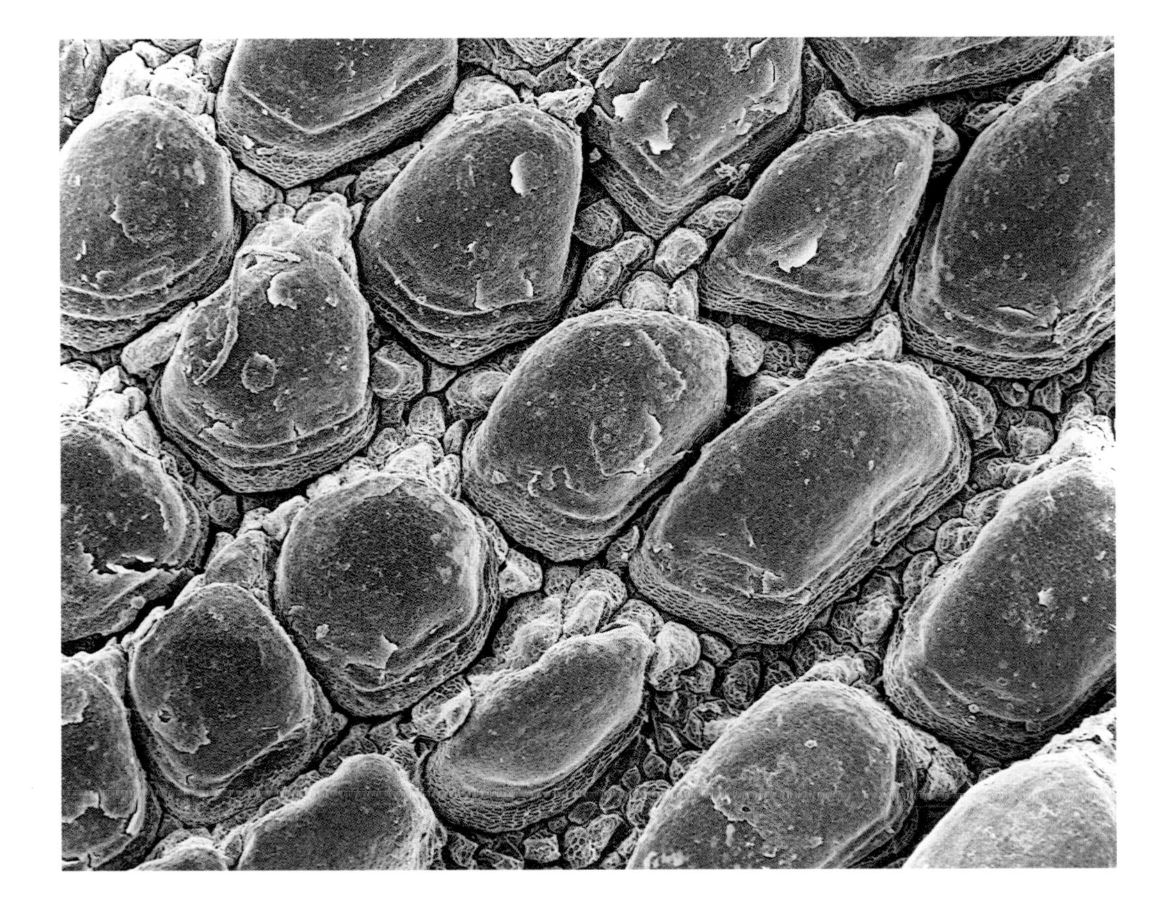

4.13 Iguana skin, x 50 (22)

4.14 A spill of coccoliths, x 9,500 (3, 17)

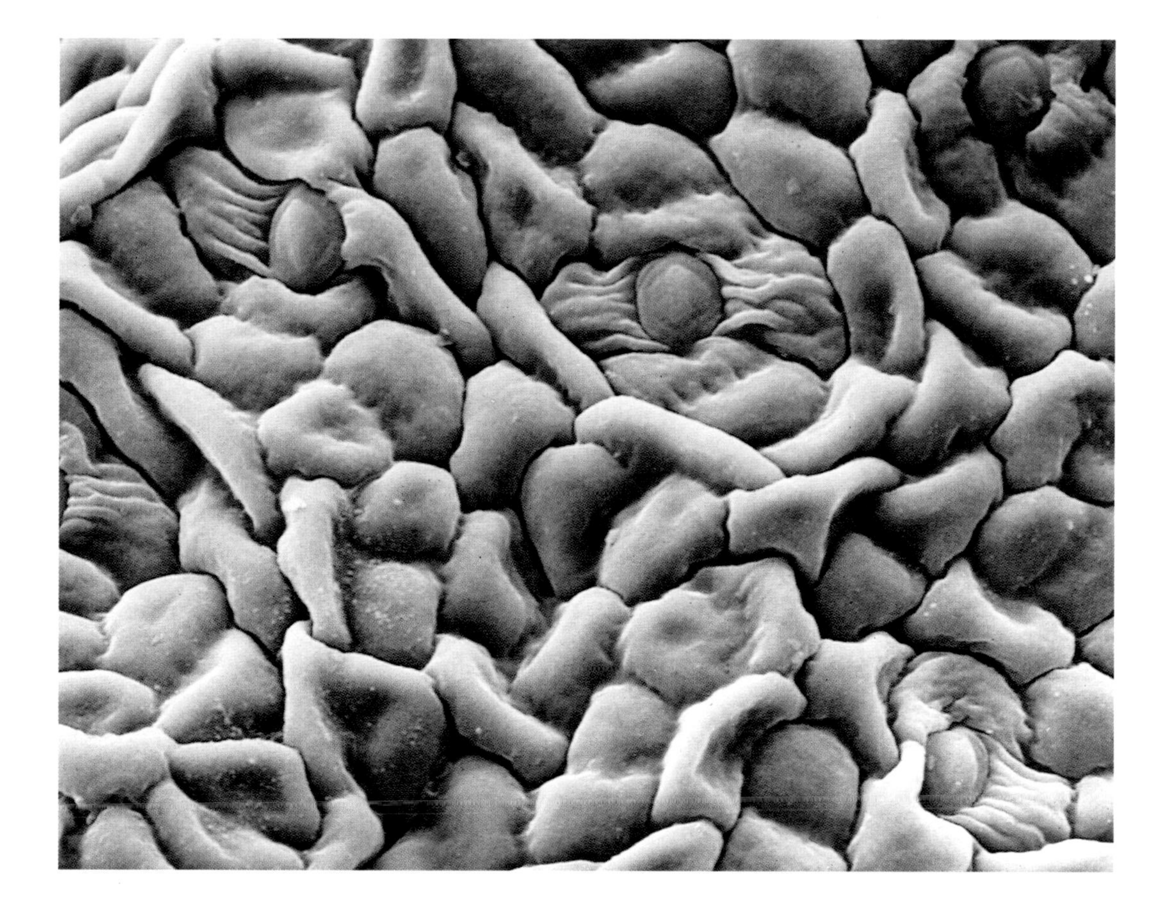

4.15 On the surface of a goldenrod leaf, x 800

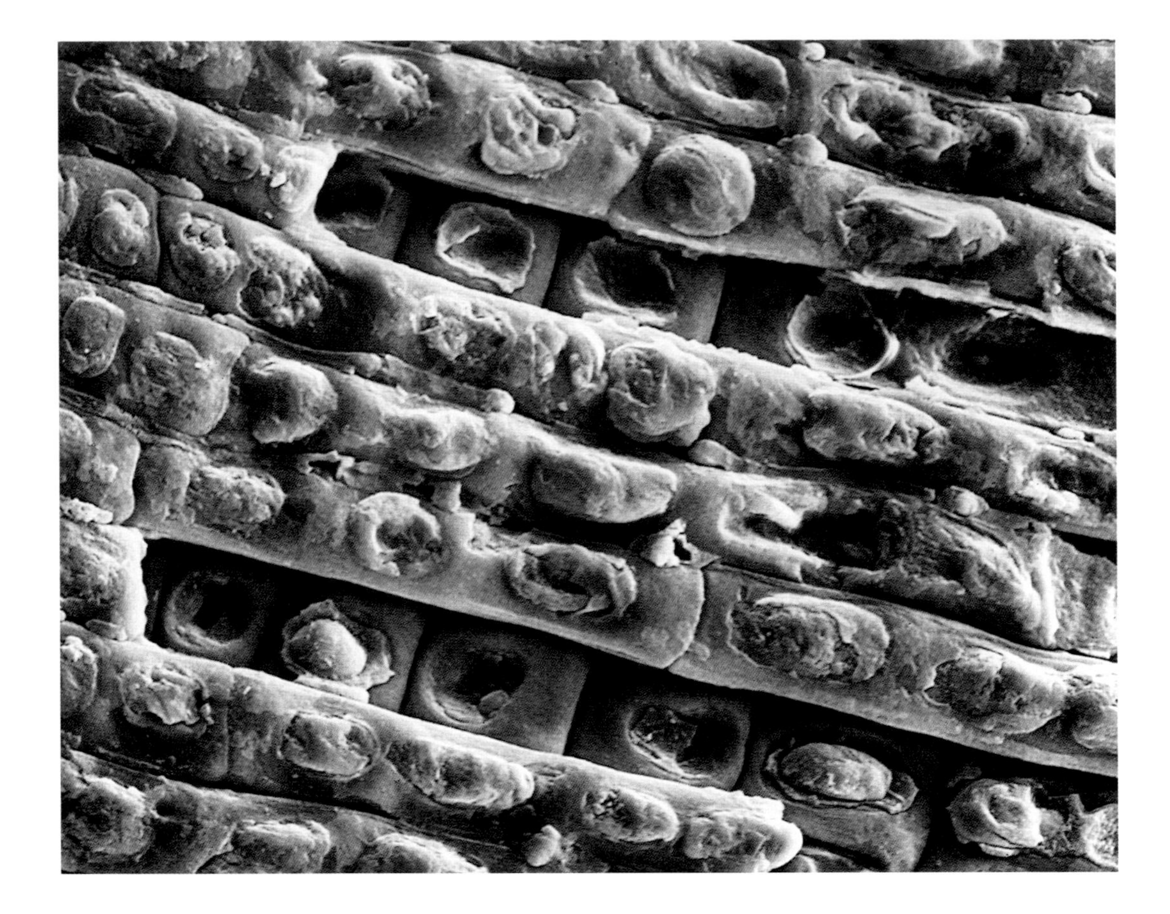

4.16 TWO SETS OF CELLS OVERLAYING EACH OTHER IN PETRIFIED WOOD: the sap in the living tree trickled through the pits and along the cells, x 800 (7)

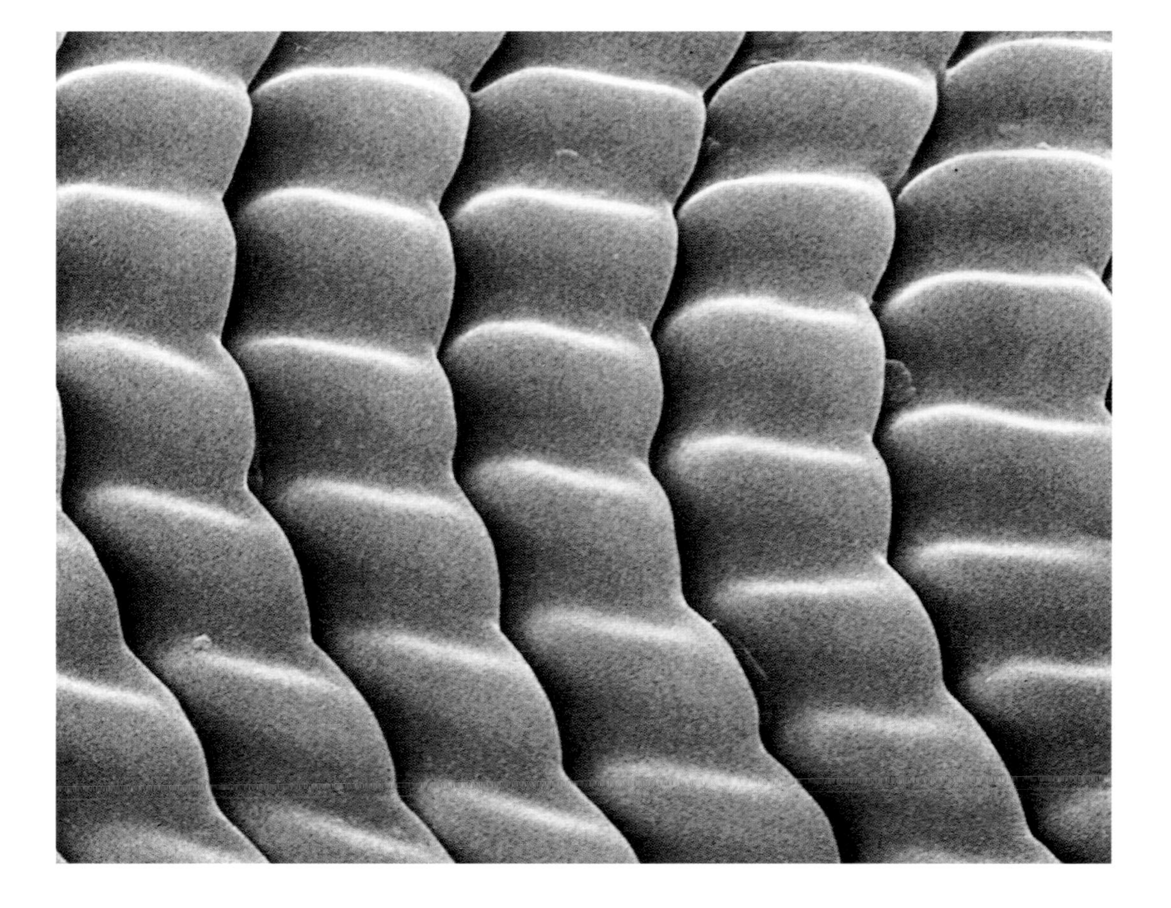

4.17 The coil of a butterfly's proboscis, x 3,500 (6)

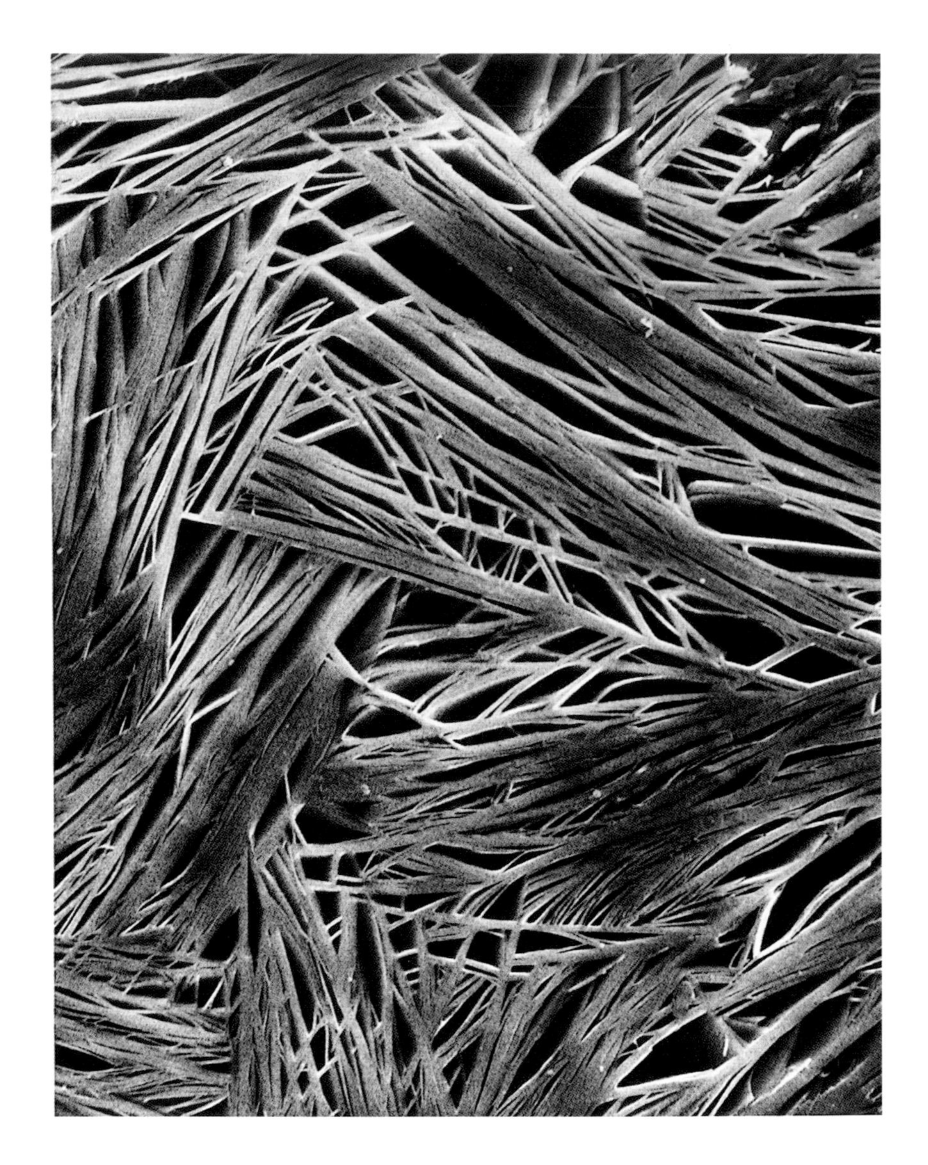

4.18 Franklin micromineral, x 2,400 (13)

4.19 Down feather, x 170

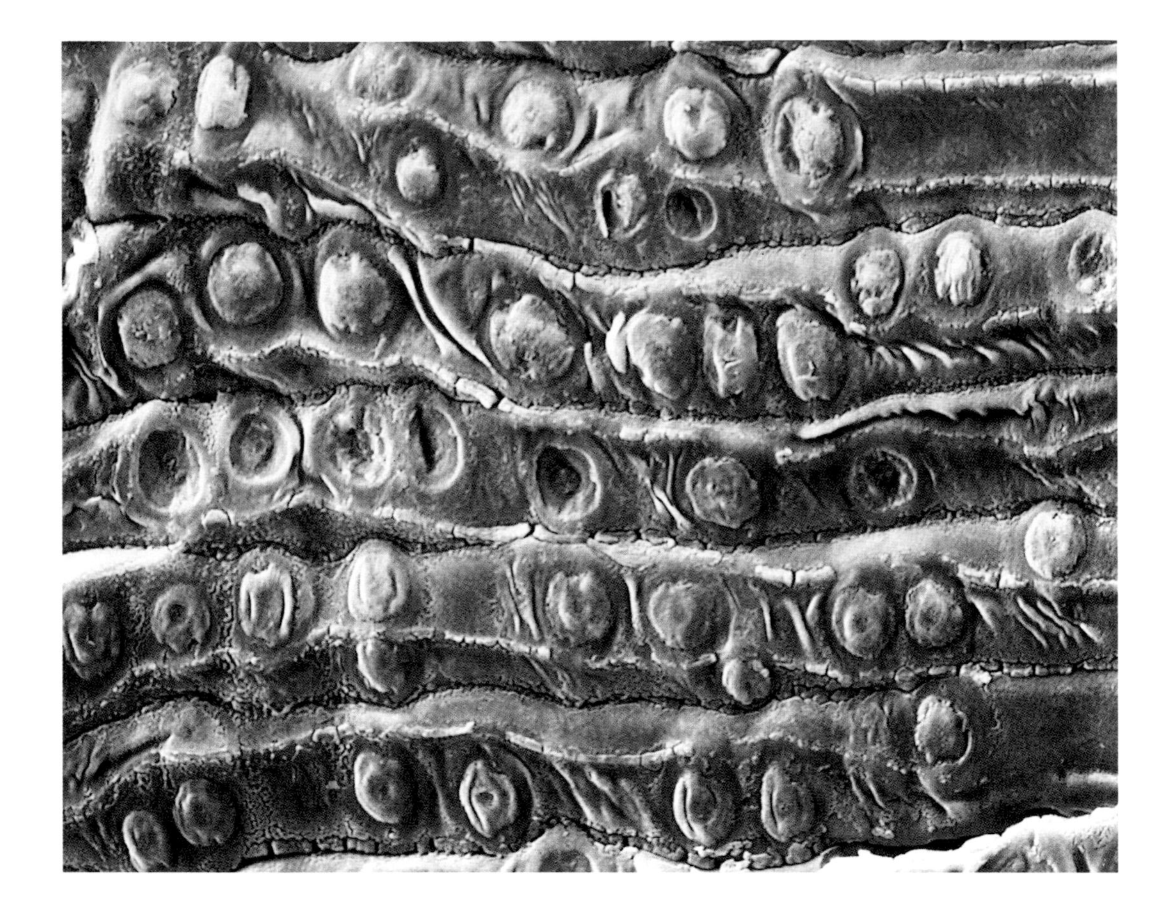

4.20 Wood from an ancient conifer, distorted in the process of petrification, x 600 (7)

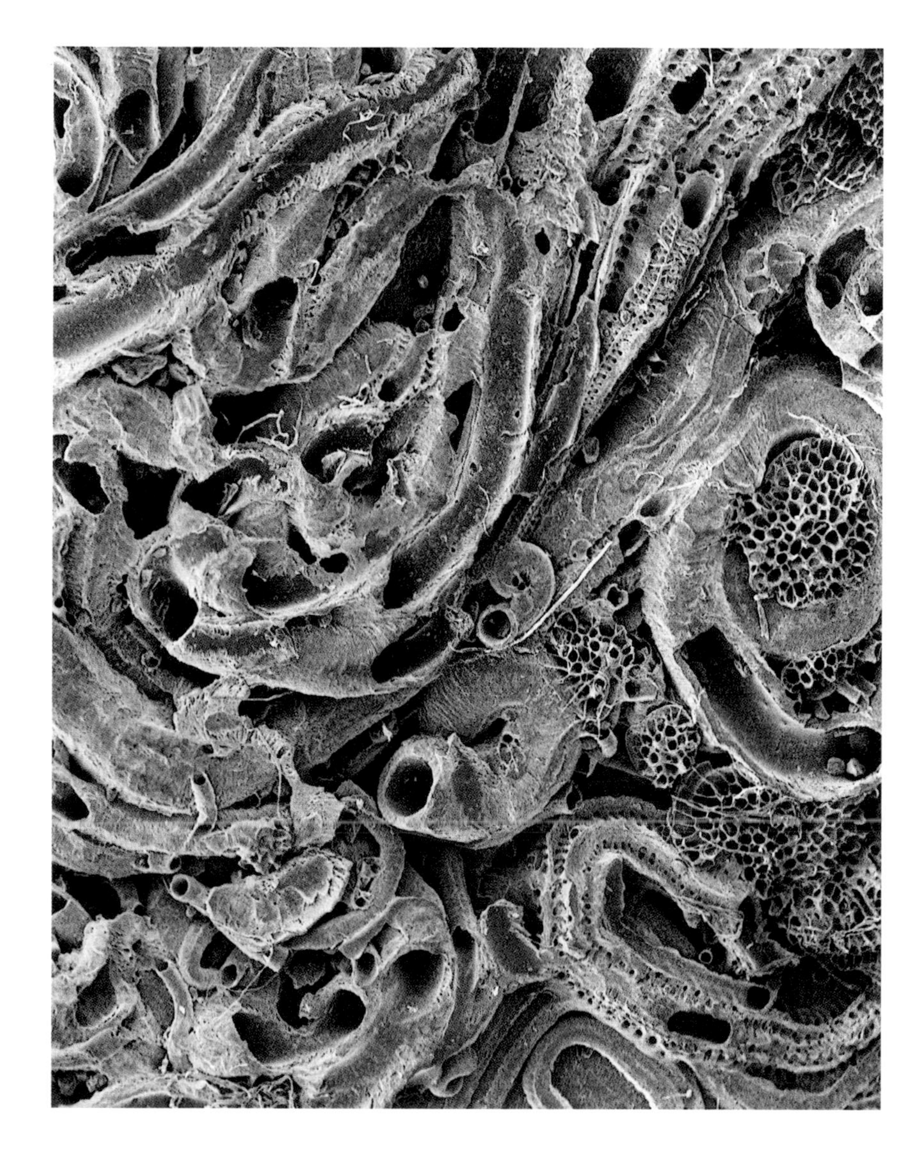

4.21 Bird's-eye view of worm tubes on a shell from Baja California, x 20 (27)

4.22 Interleaved coccoliths, protecting the cell inside, x 26,000 (3, 17)

4.23 VAN GOGH REVISITED: intimately mixed minerals, formed when Norway and Greenland parted company about 55 million years ago, x 2,800 (37)

4.24 Laboratory-grown crystals from an experiment investigating the formation of kidney stones, x 4,300 (38)

4.25 Industrial filter, x 22,000 (8)

4.26 MOSQUITO WING SERIES: surface detail, x 4,700 (6)

4.27 Scales on the wing of a Painted Lady butterfly, x 650 (6)

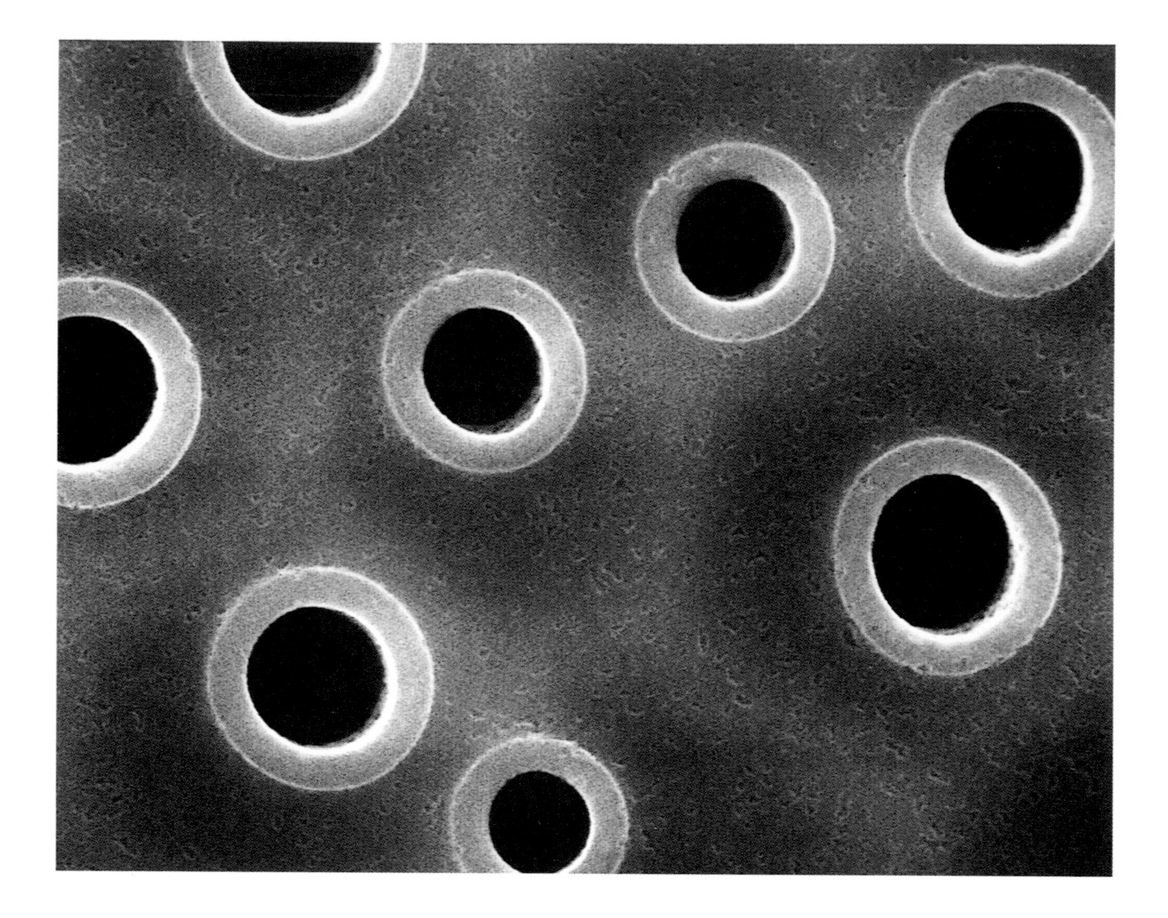

4.28 Detail of a diatom from the undersea Hendrickson Canyon, which cuts through the continental slope off the coast of eastern North America, x 14,000 (2, 39)

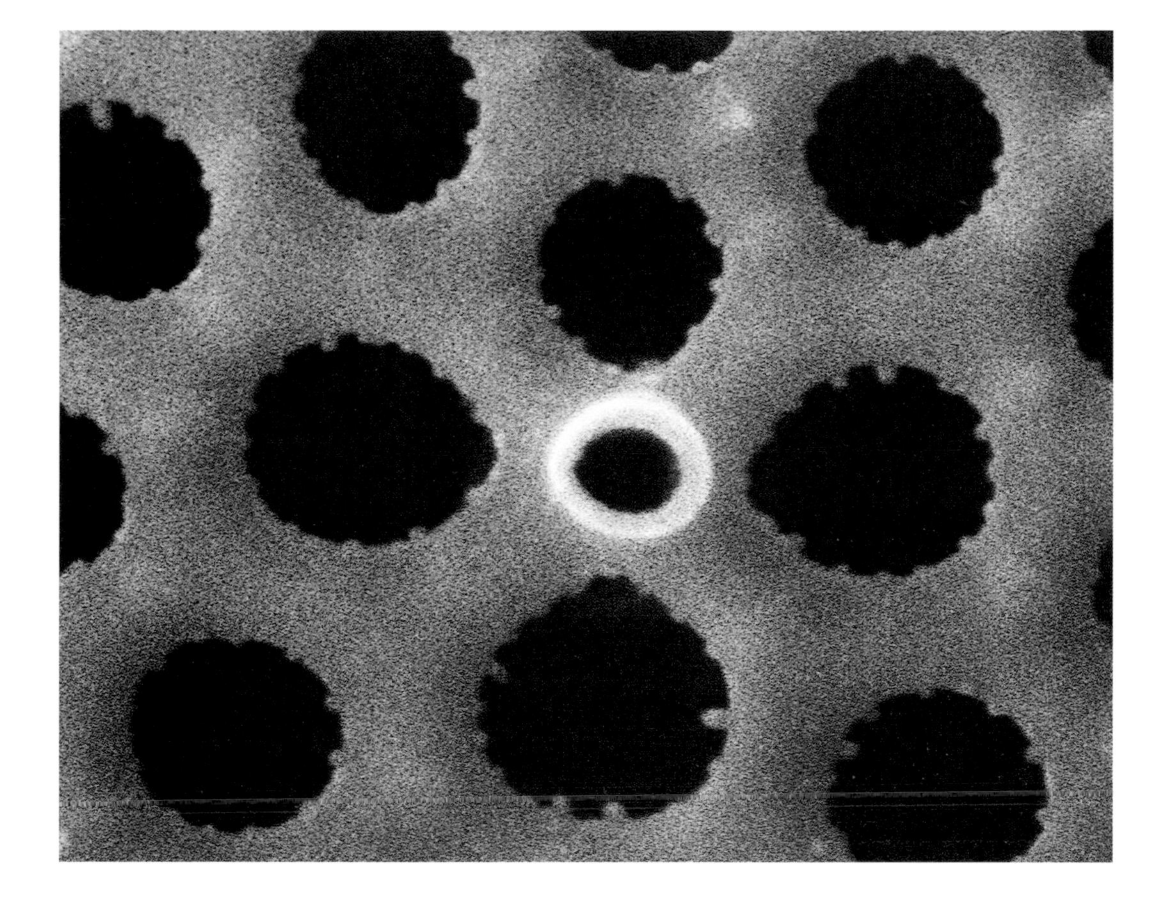

4.29 CAUGHT IN THE ACT: detail of a diatom that has begun to dissolve; its chemicals will someday return in chalk in the walls of the Hendrickson Canyon, x 55,000 (2, 39)

4.30 On the inner surface of a robin's eggshell, x 350 (40)

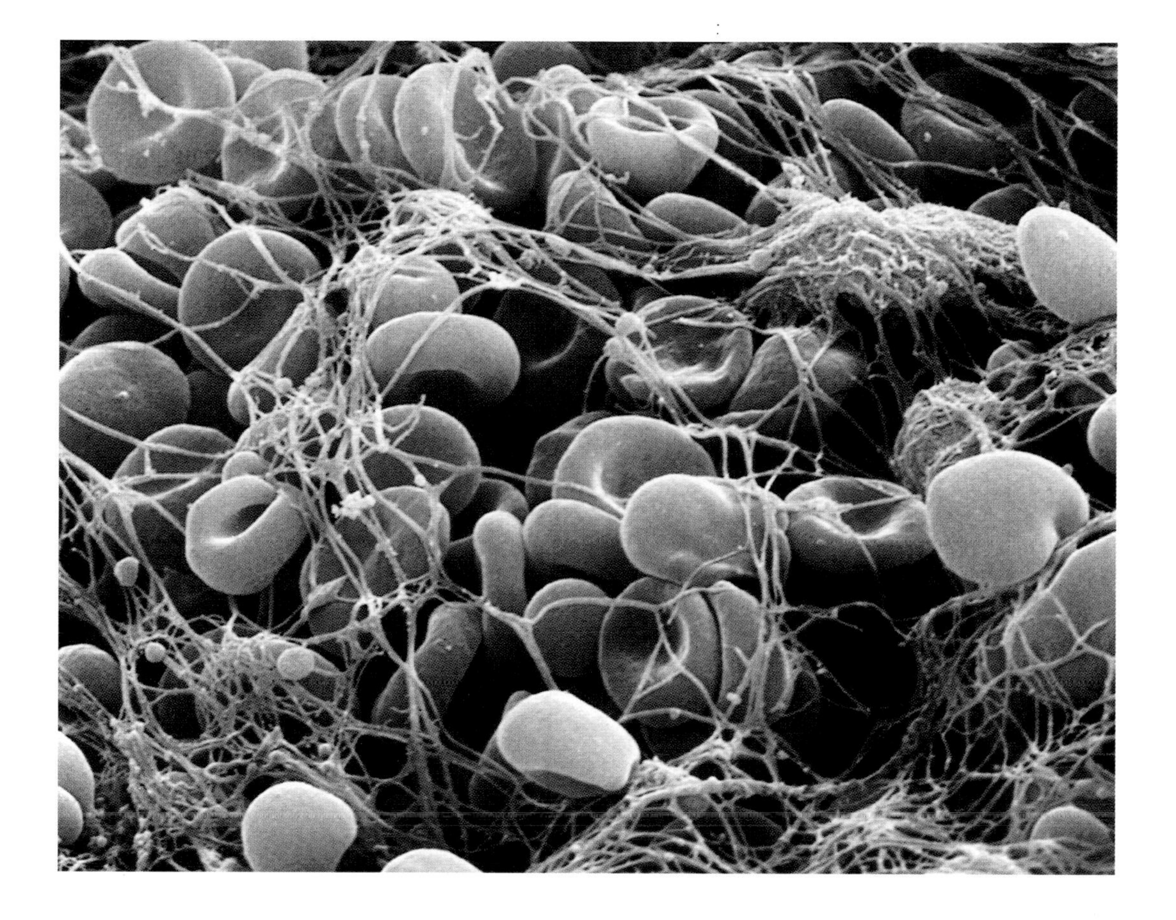

4.31 Arterial blood clot, x 4,200 (33)

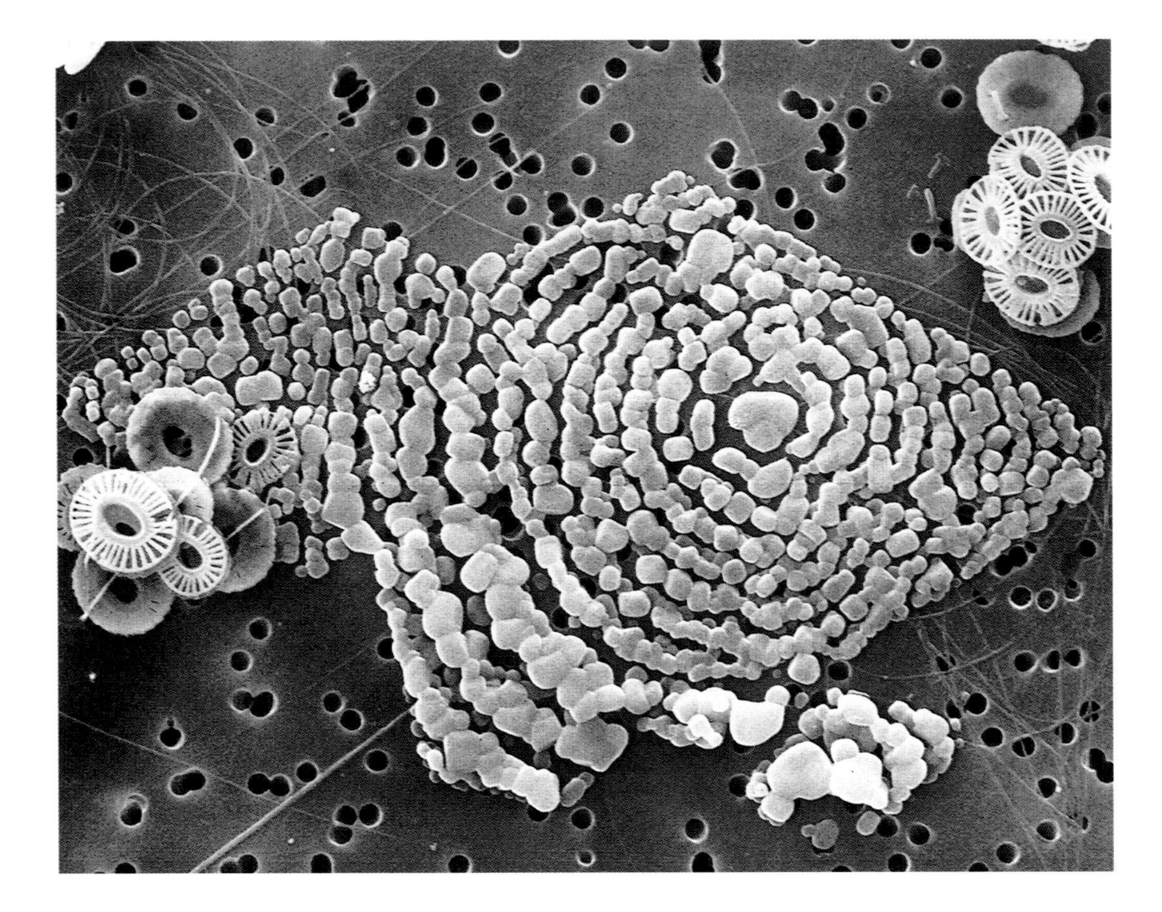

4.32 Two overlapping depositions of sea salt, dried on a filter that was used to catch microplankton living in the sea near Iceland, x 4,000 (41)

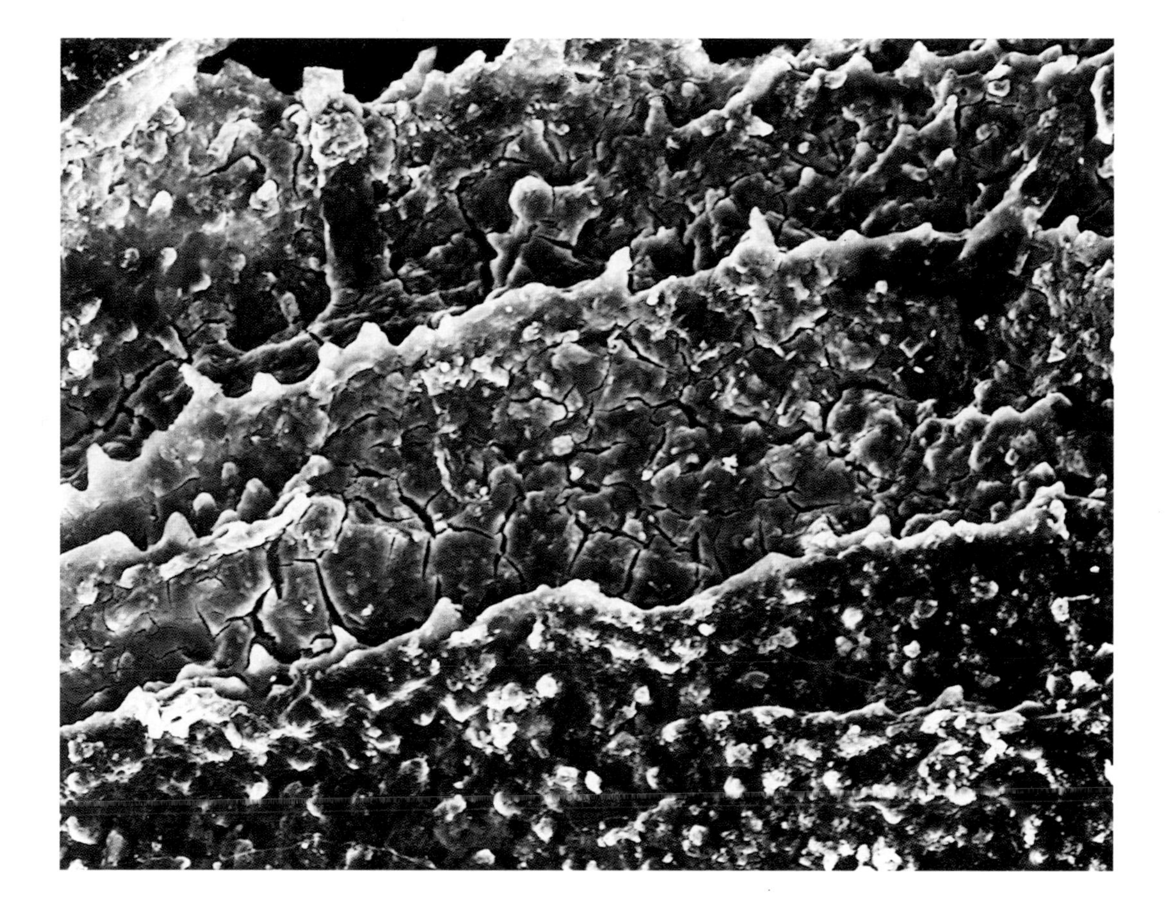

4.33 FOOD FOR DINOSAURS: surface detail on a 220-million-year-old conifer leaf from present-day Pennsylvania, x 300 (42)

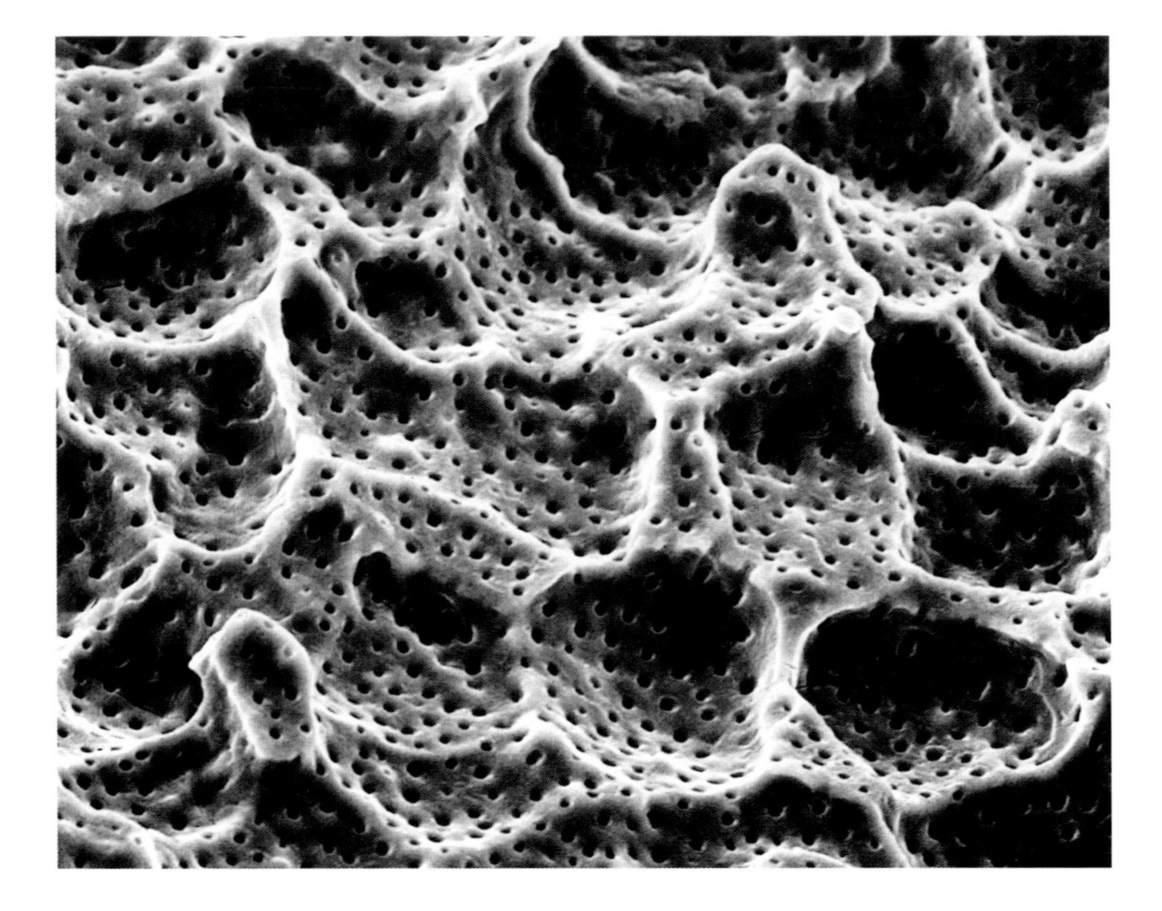

4.34 CAUGHT IN THE ACT: detail of a baby tooth, showing the craters where specialized cells sat eating dentin to loosen the tooth (and the tubules that connected the nerve with the surface), x 700 (43)

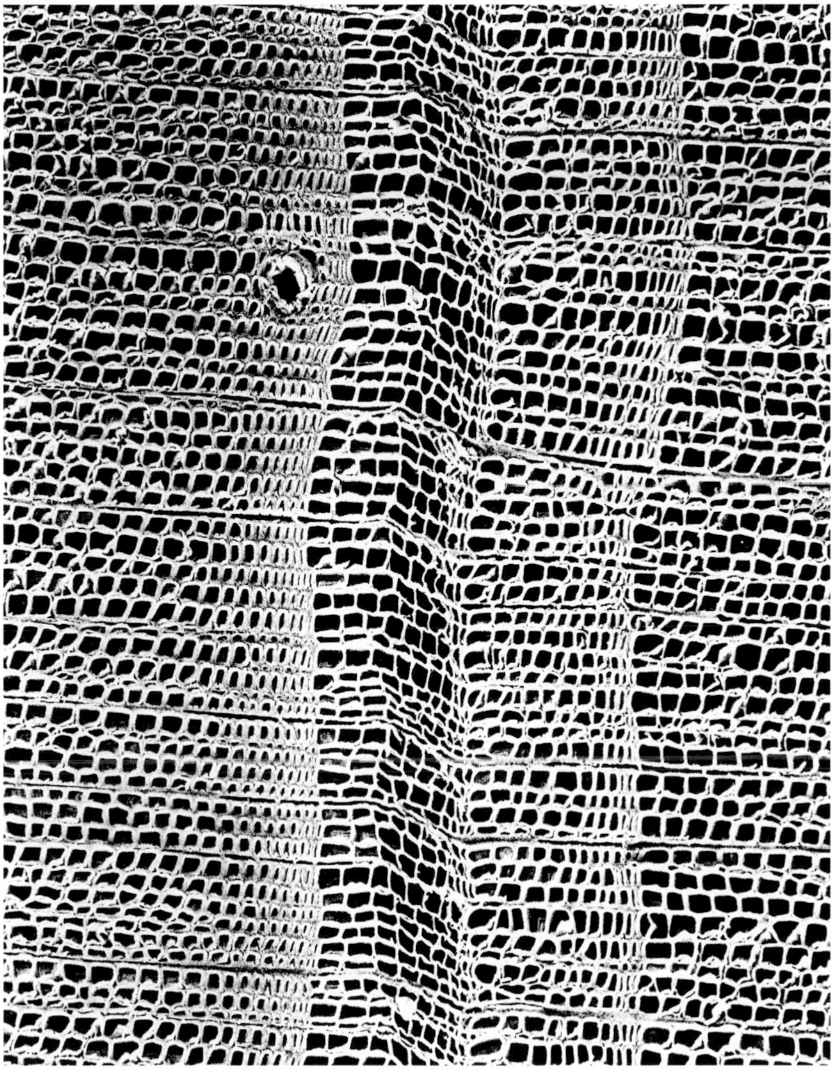

4.35 TREE RINGS FROM A CONIFER LIVING IN THE NORTHWEST TERRITORIES: the deformed narrow rings in the center reflect two unusually cold summers that almost killed the tree in 1641 and 1642, x 115 (35)

4.36 LABORATORY CRYSTALLIZATION OF SILICATE MAGMA: a petrology experiment, designed to understand the natural occurrences of similar ancient lavas, x 3,500 (44)

4.37 The compound eye of a fly, x 1,500 (45)

4.38 Surface detail of a human liver stone, x 5,800 (46)

4.39 Franklin micromineral, x 650 (13)

5.1

5 · PORTRAITS

5.2 Coccolith snared in a filter, x 10,300 (3, 17)

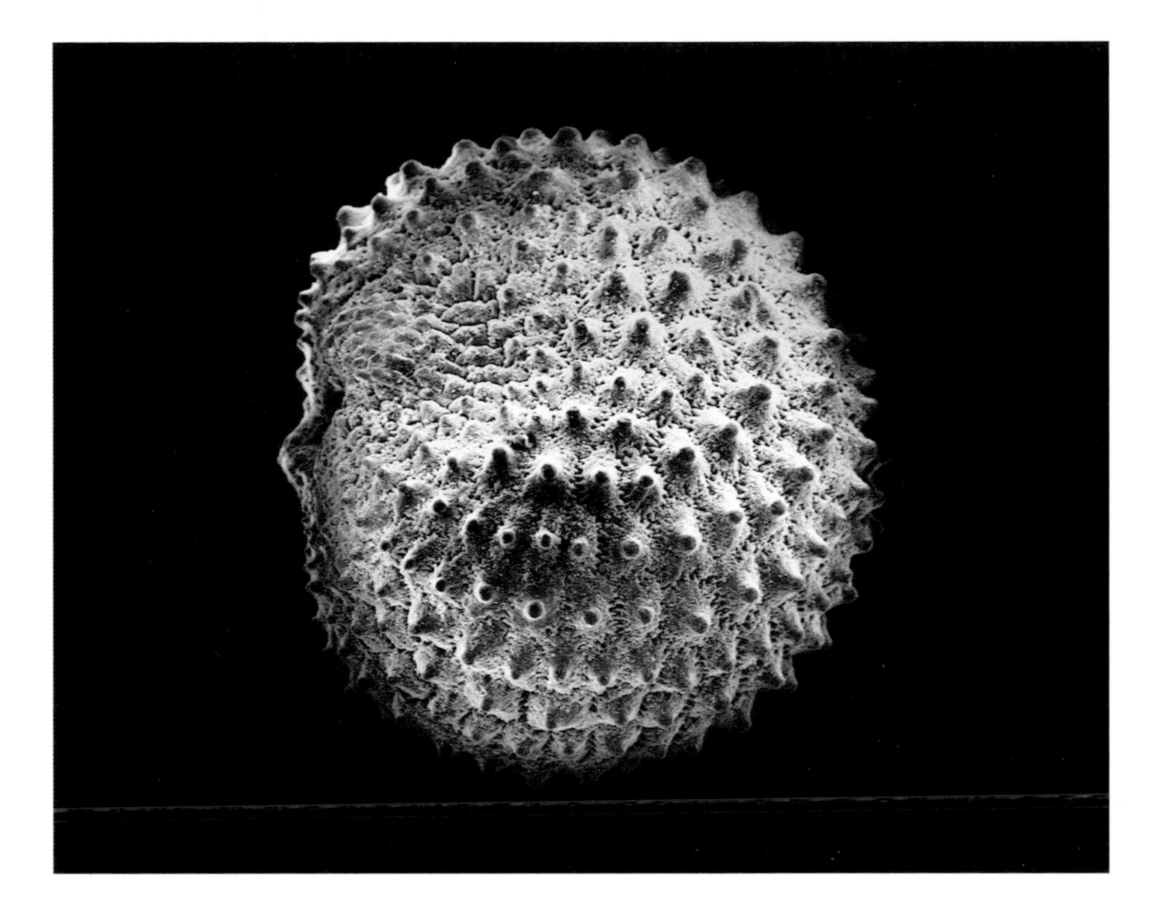

5.3 Seed of a Drooping Catchfly from Holland, x 65 (48)

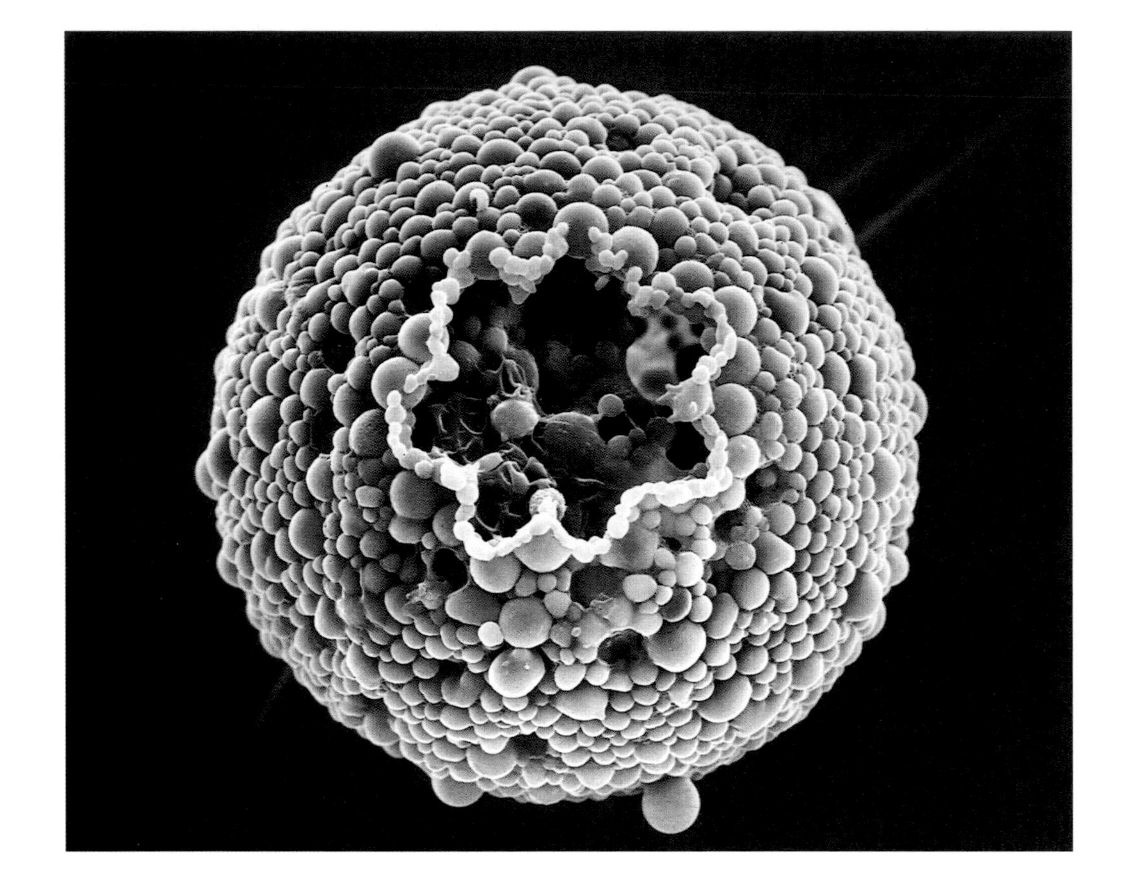

5.4 A shelled amoeba, x 1,000 (49)

5.5 Human pancreatic stone, x 35 (50)

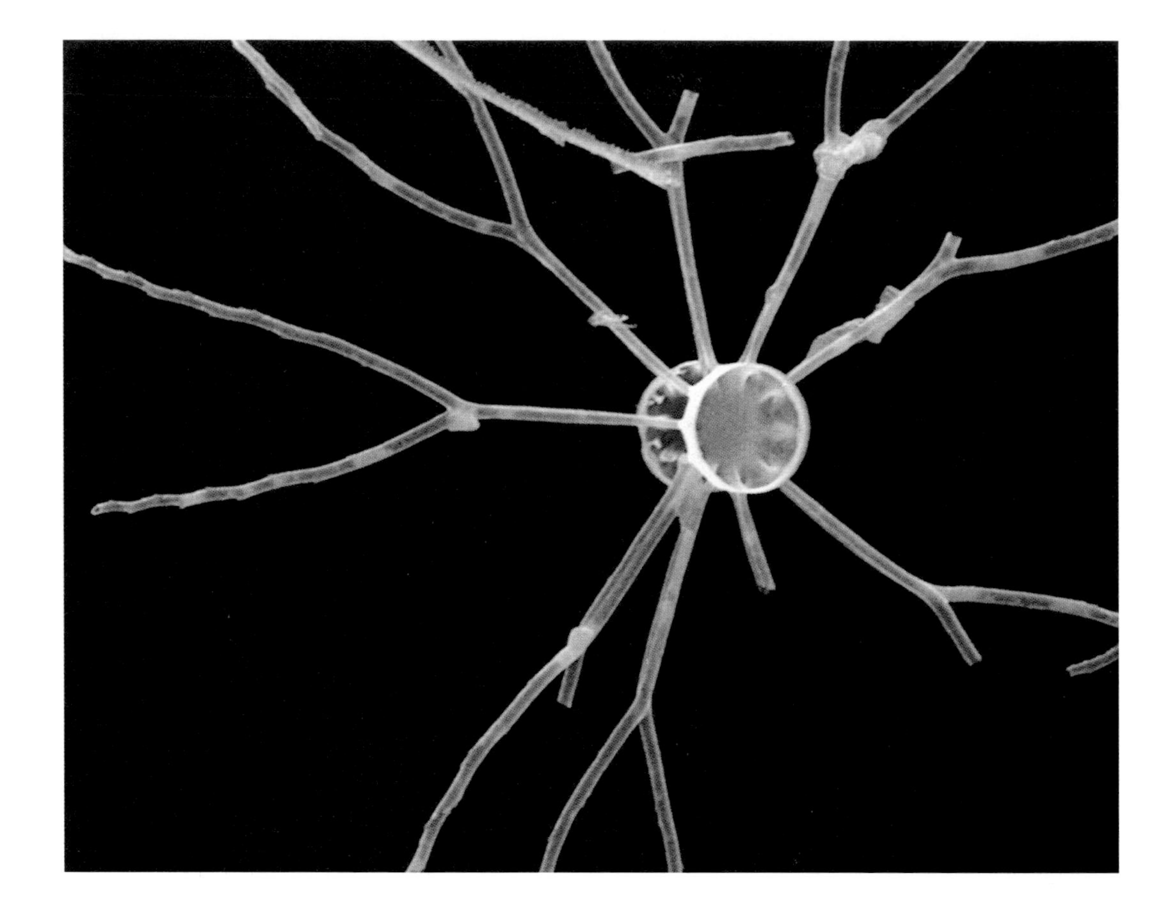

5.6 Caribbean diatom, x 1,850 (2, 51)

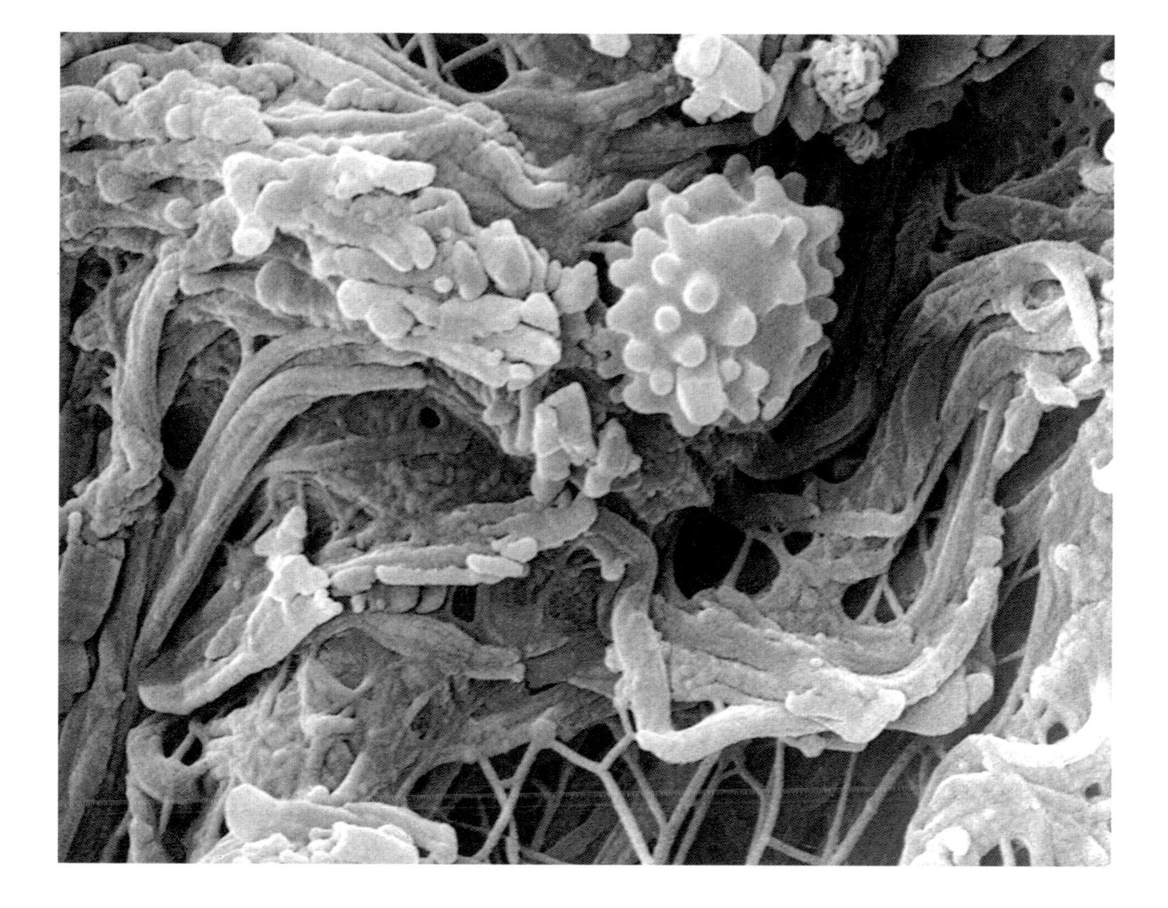

5.7 White blood cell in connective tissue from a human heart, x 6,400 (20)

5.8 Caribbean radiolarian, x 450 (1, 51)

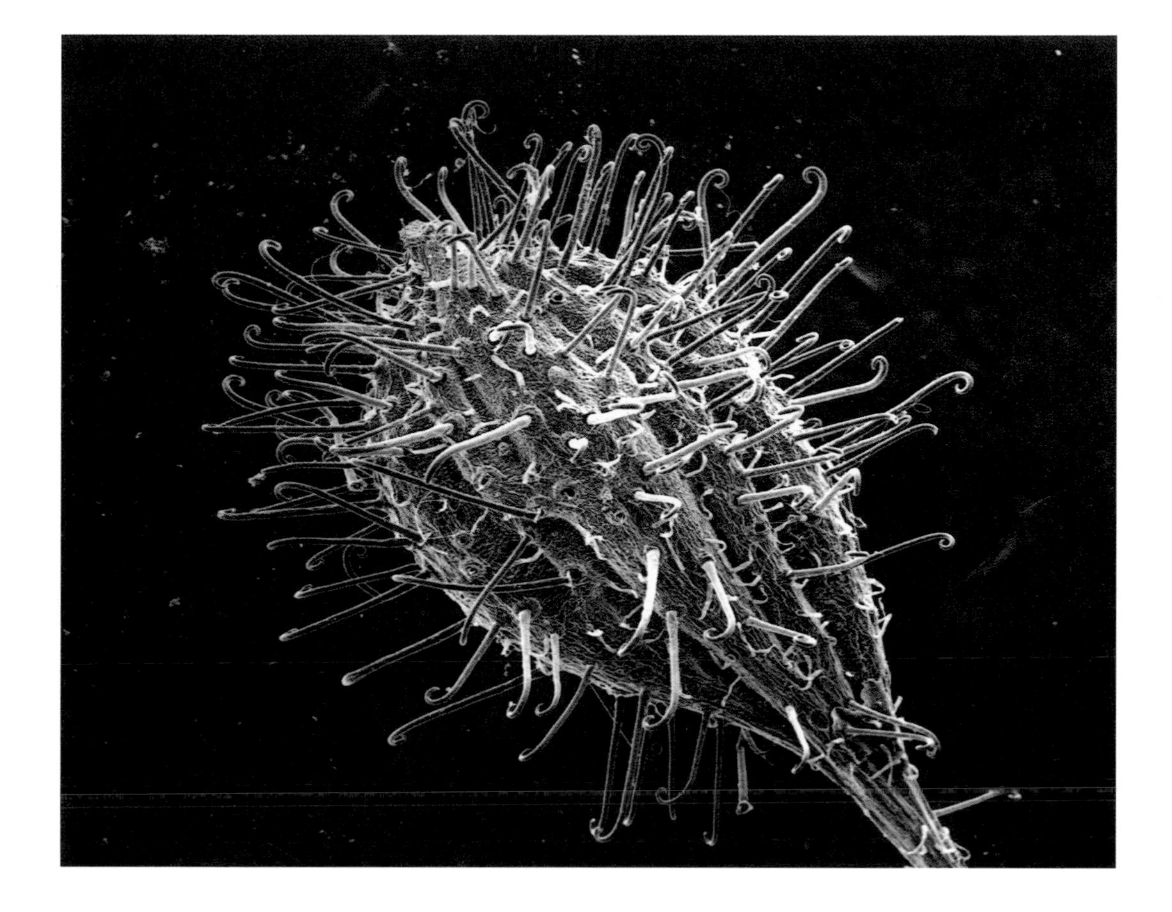

5.9 ENCHANTER'S NIGHTSHADE: a seed from West Virginia, x 23 (48)

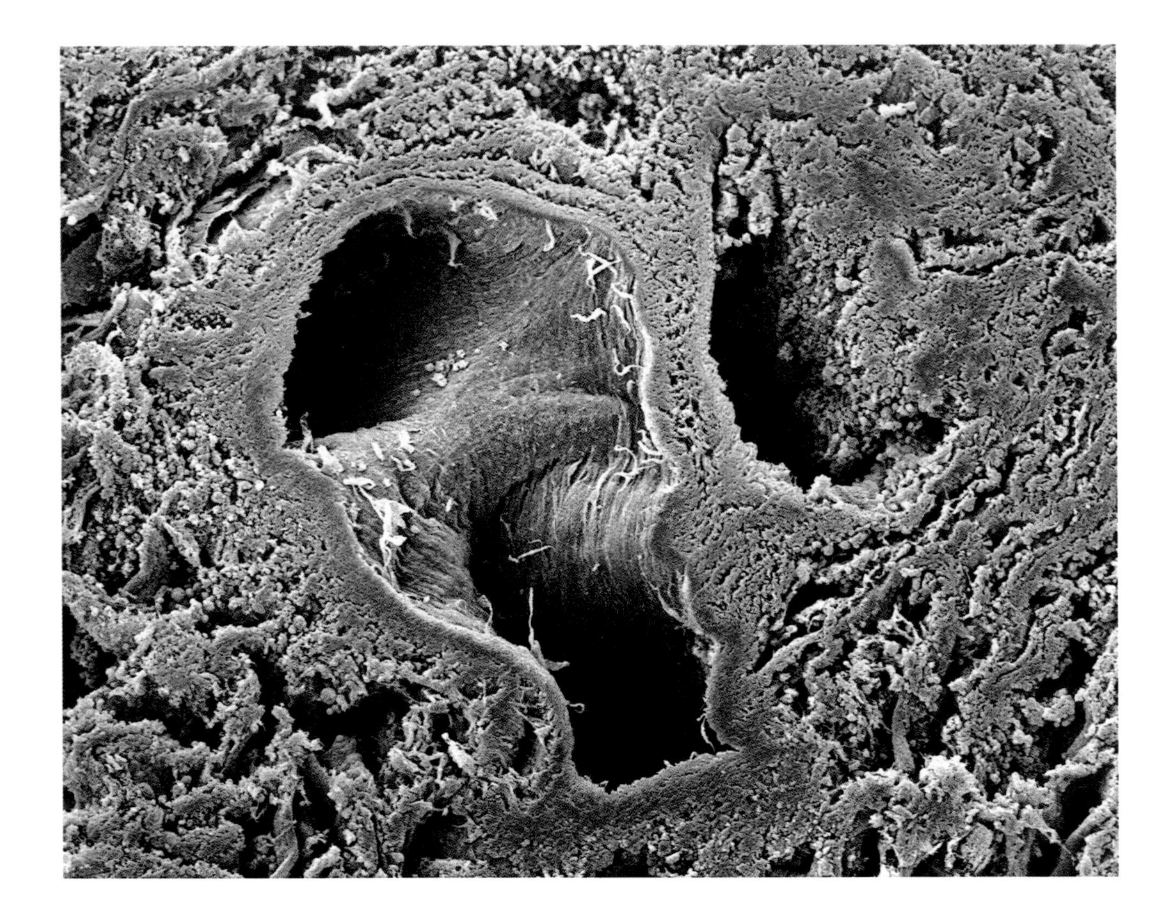

5.10 CAUGHT IN THE ACT: human pulmonary artery, branching (and its companion bronchiole), x 1,200 (20)

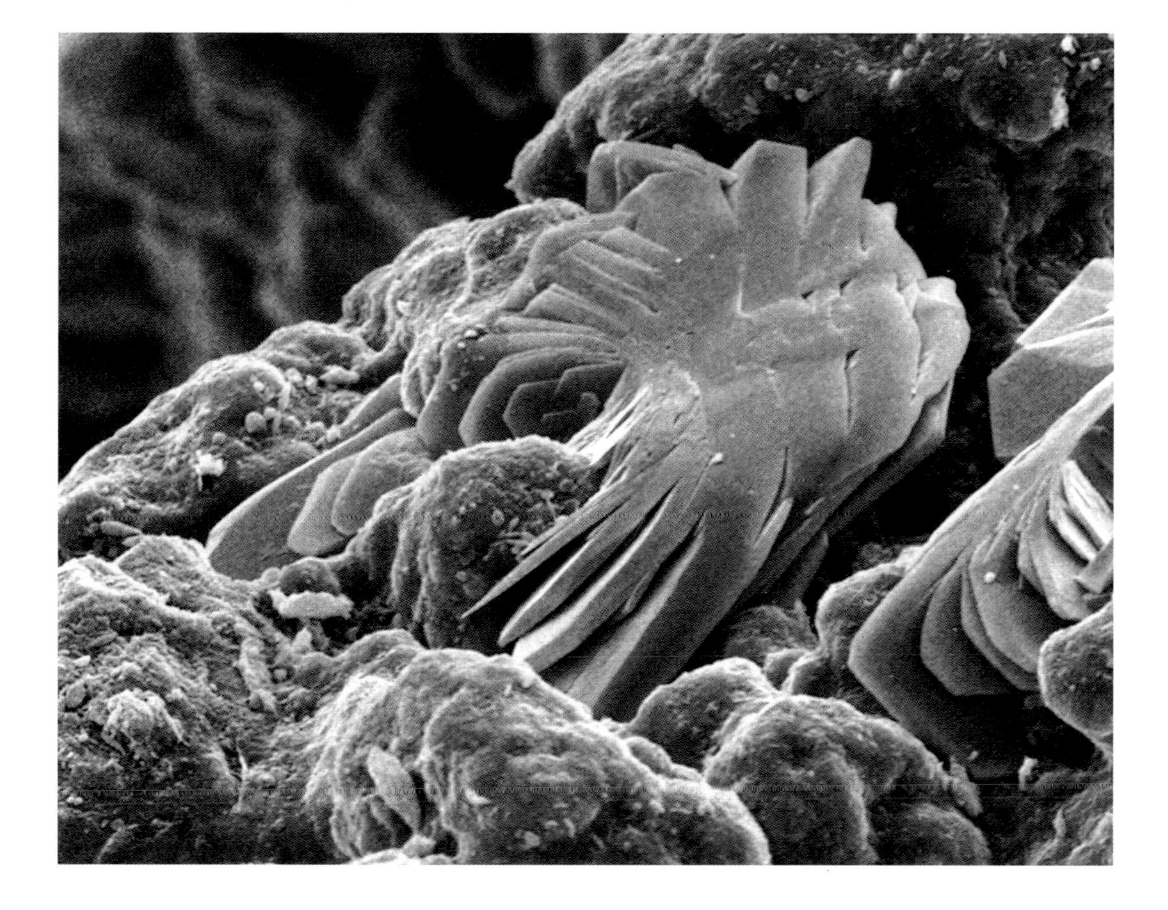

5.11 A crystal of gypsum from an ancient shoreline of Pyramid Lake, Nevada, x 950 (52)

5.12 Synthetic kidney stone crystal, x 4,300 (38)

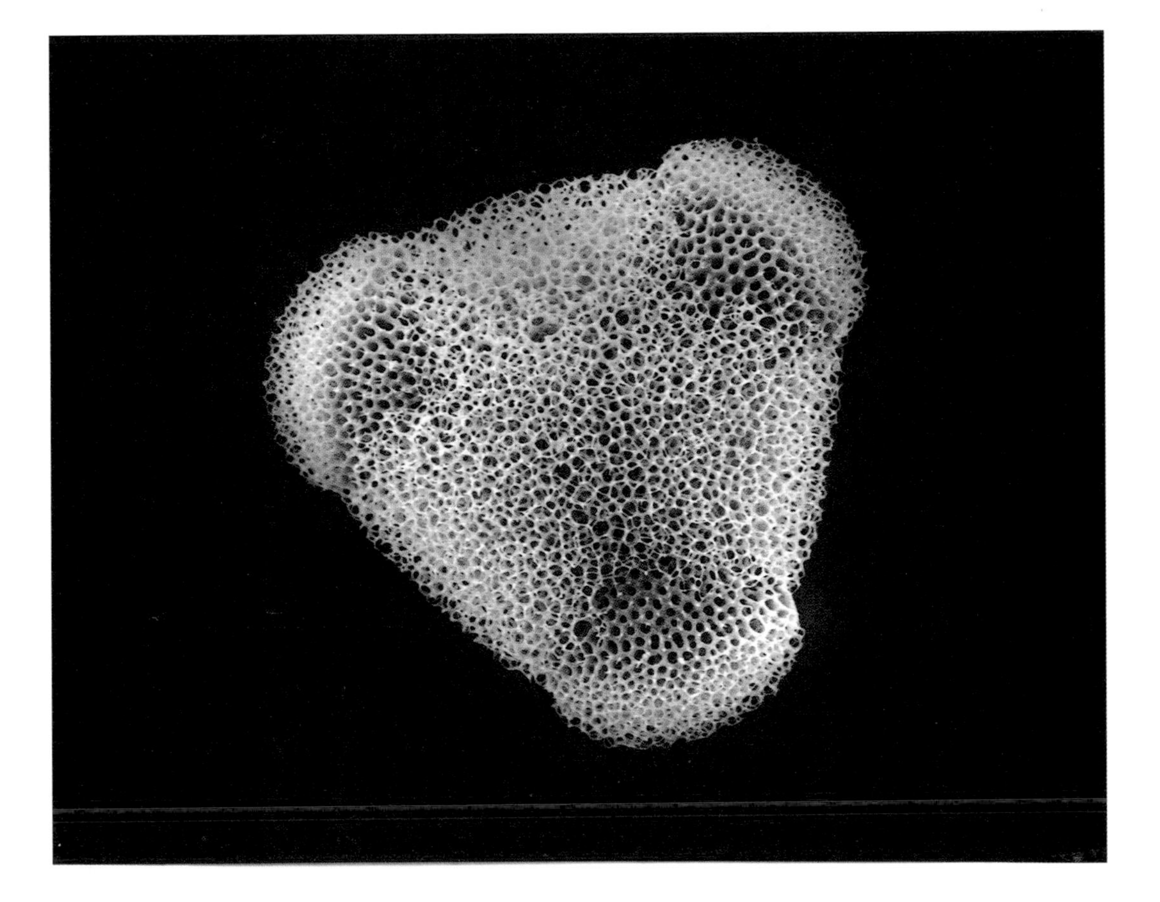

5.13 Caribbean radiolarian, x 250 (1, 51)

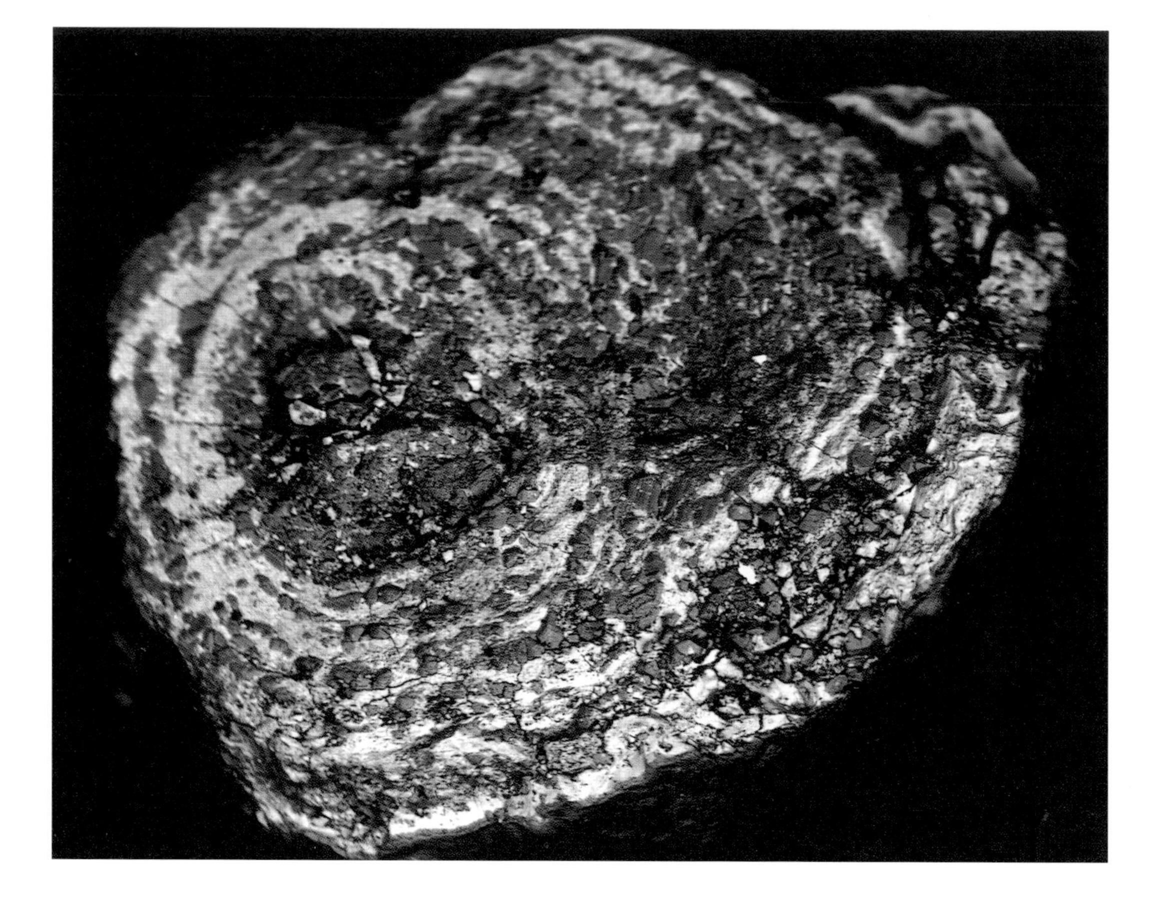

5.14 Chemical image of a human kidney stone, showing that it grew like a pearl, x 22 (38)

5.15 Shards of ash from a volcanic eruption that killed ancient animals gathered at a watering hole ten million years ago, x 700 (53)

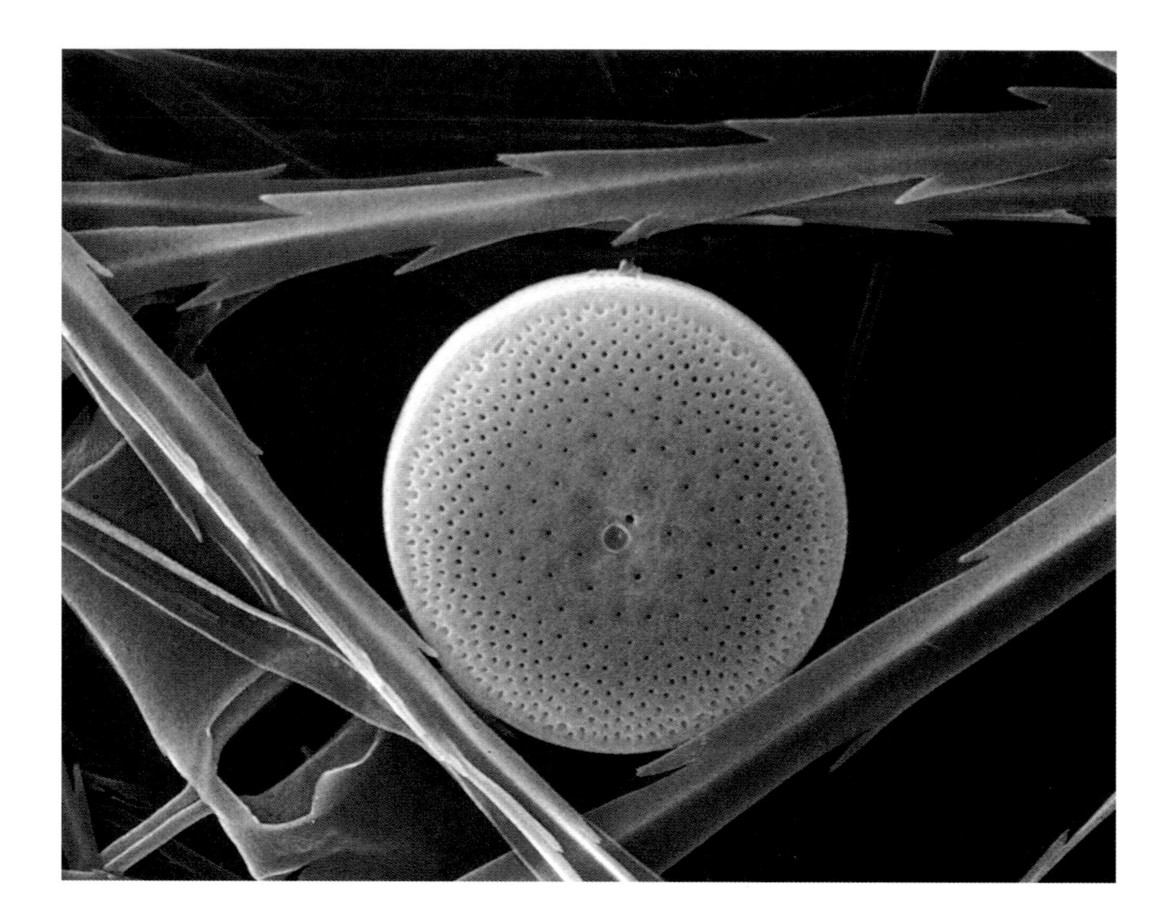

5.16 Antarctic diatom, resting among parts of other diatoms, x 4,300 (2, 54)

5.17 Side view of a diatom, x 7,200 (2, 17)

5.18 Dandelion seeds, revealed within the flower, x 90

5.19 Tropical coccolithophorid, x 11,000 (3, 17)

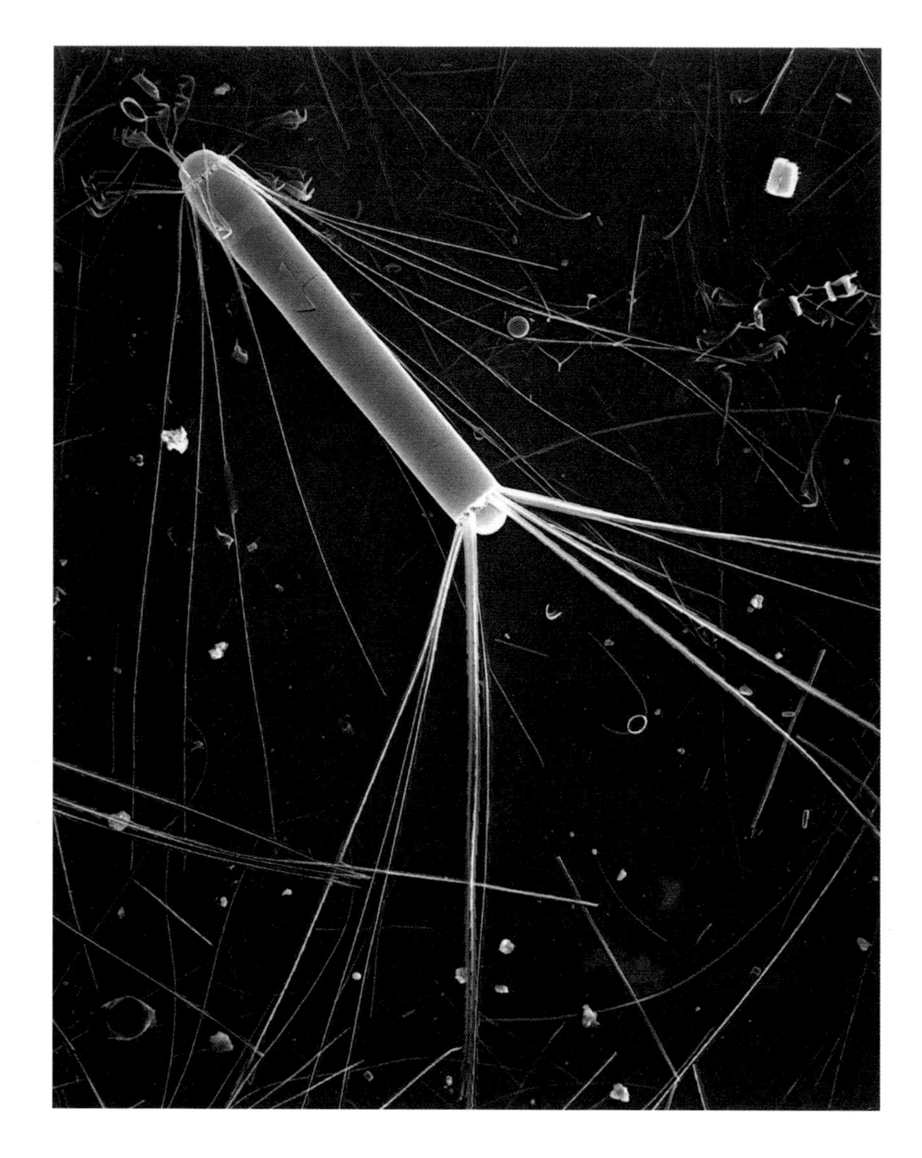

5.20 Pennate diatom, x 250 (2, 54)

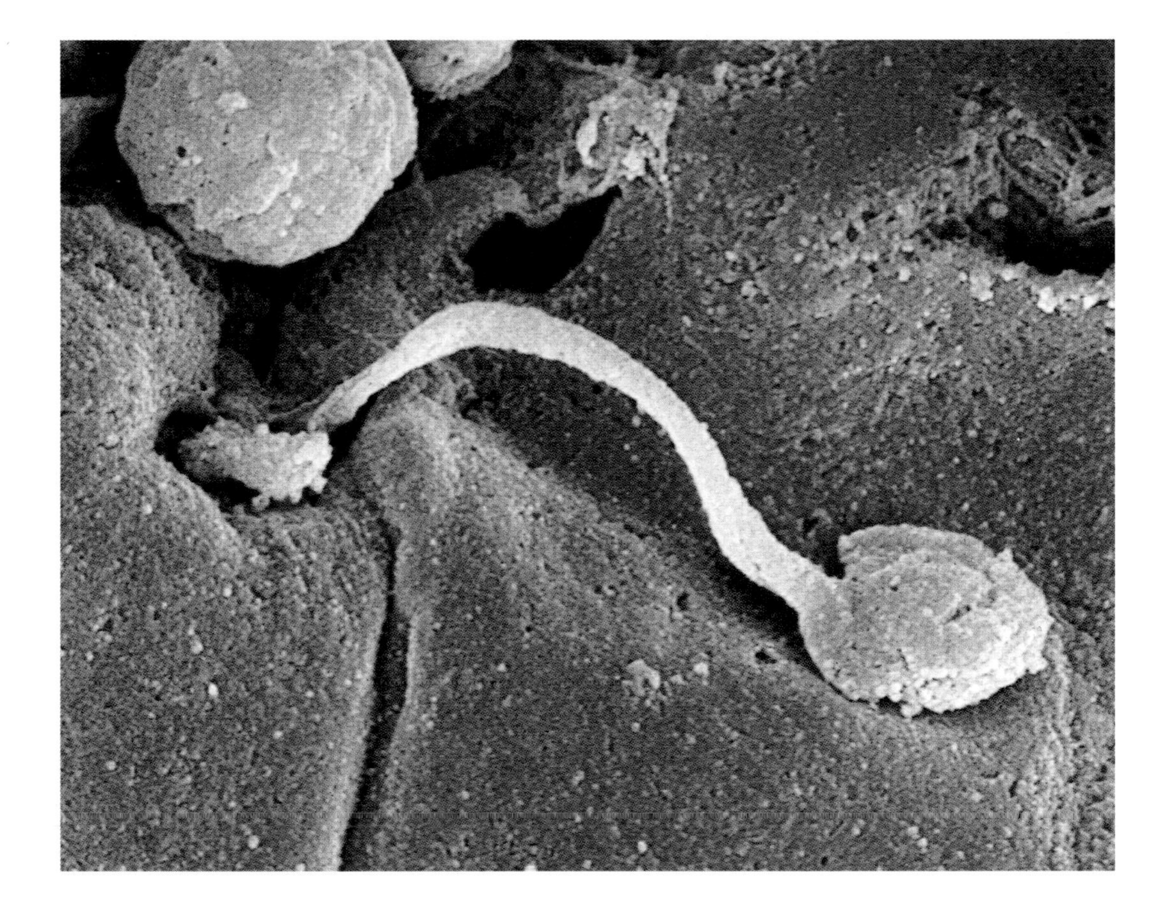

5.21 CAUGHT IN THE ACT: "walking macrophage," a white blood cell probing an air sac while cleaning a human lung with pneumonia, x 5,000 (20)

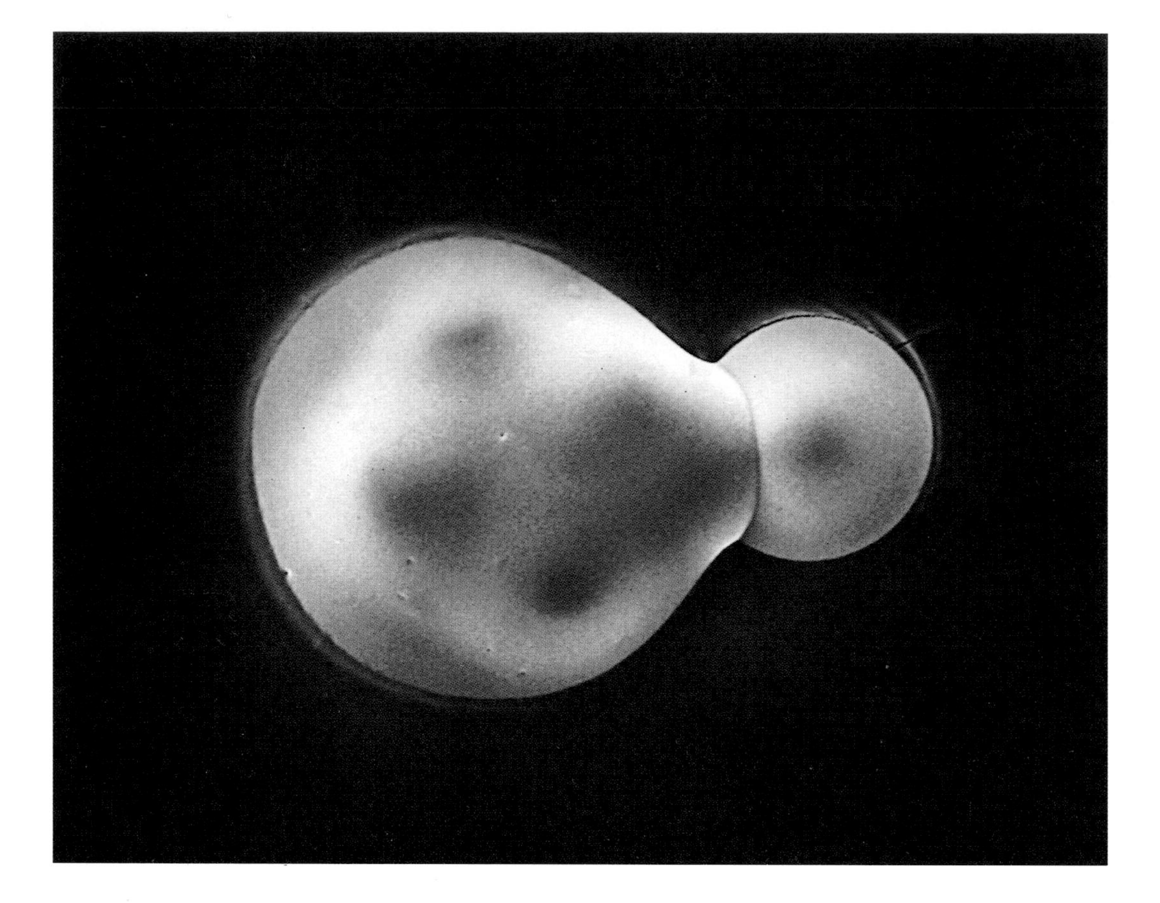

5.22 Protein from the biopsy of a human thyroid tumor, x 2,800 (55)

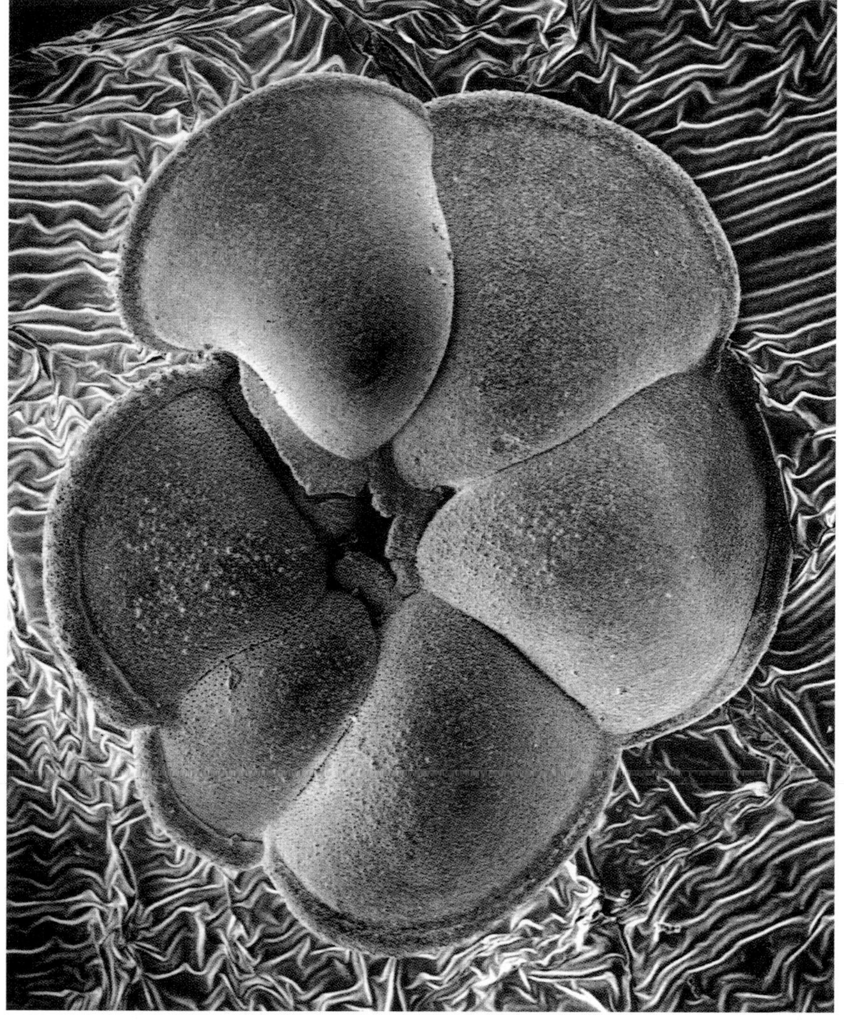

5.23 Fossil foraminifer, x 100 (4, 56)

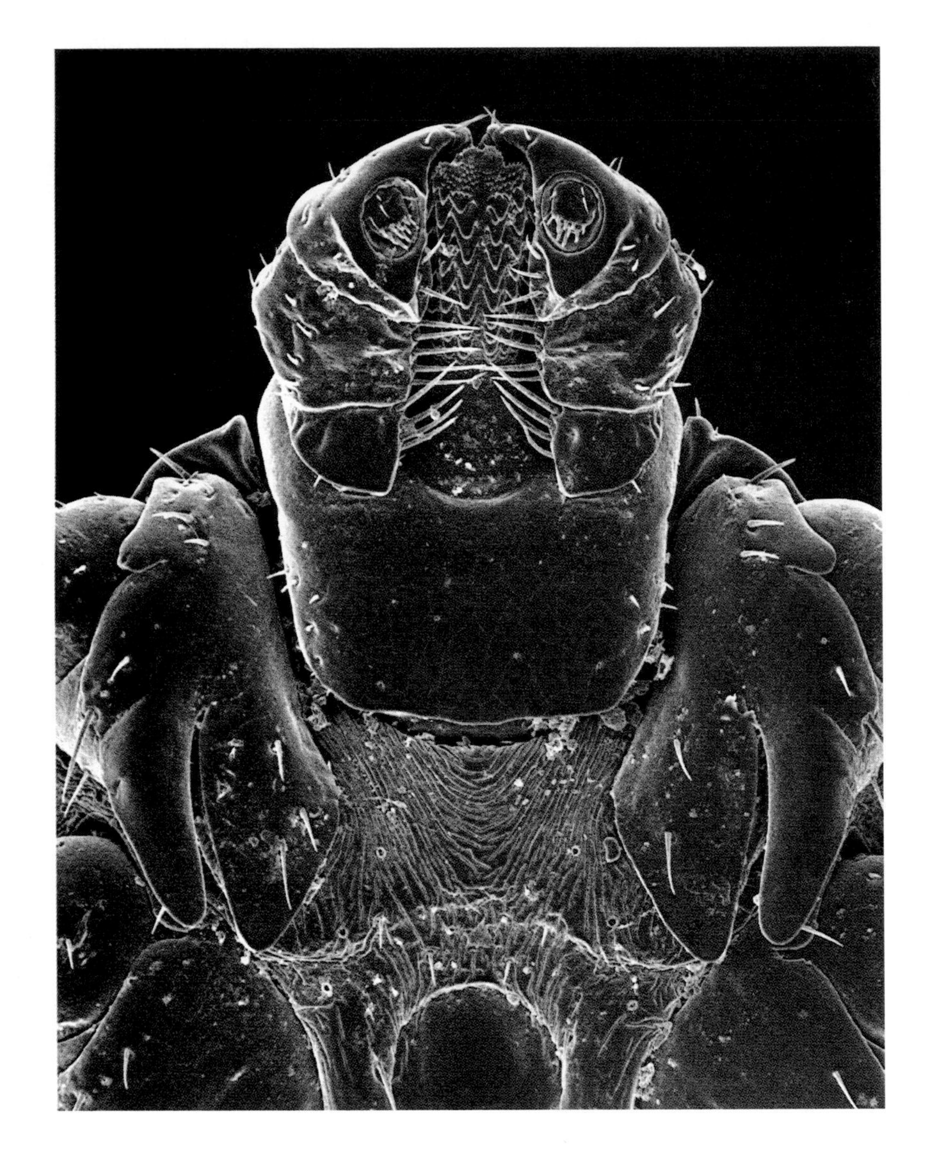

5.24 Wood tick, x 85 (6)

5.25 Plumed male mosquito, x 100 (6)

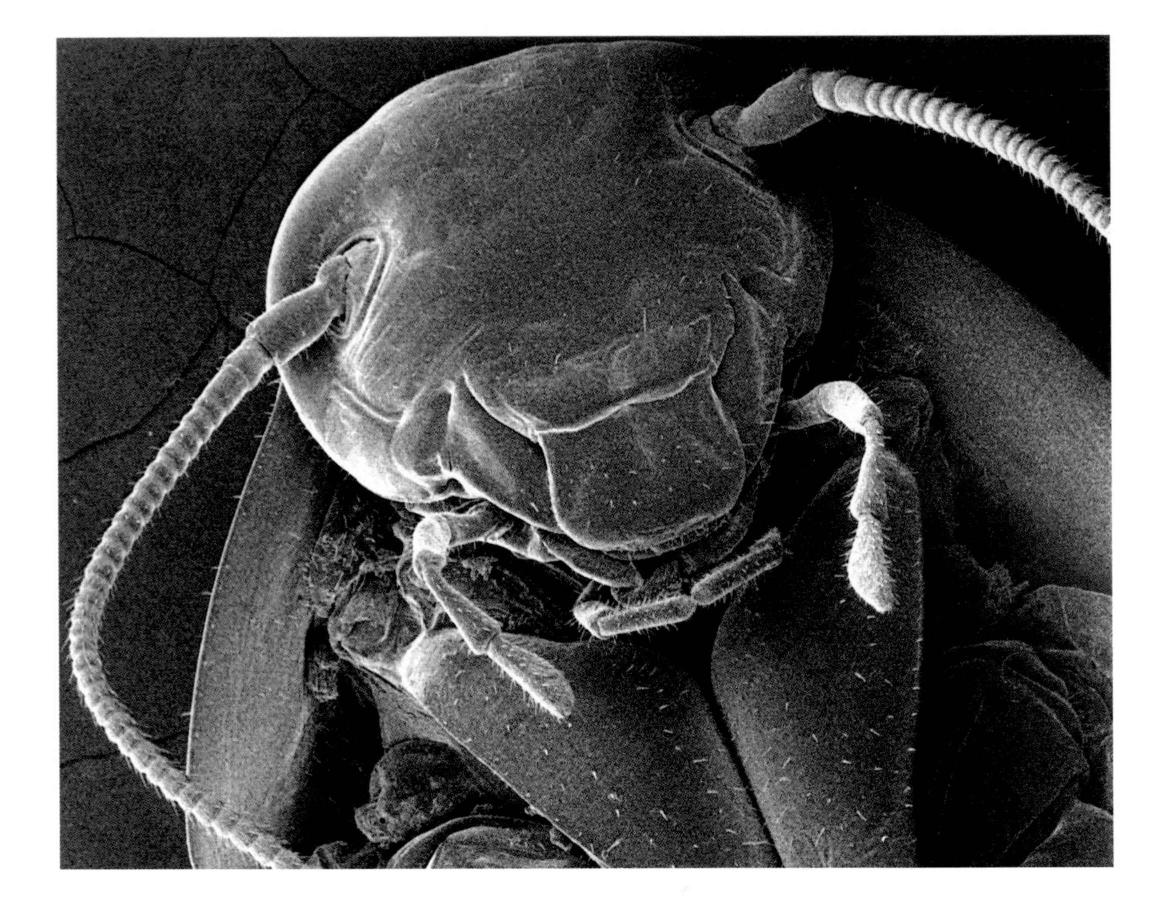

5.26 Cockroach, x 14 (6)

5.27 Millipede, x 22 (6)

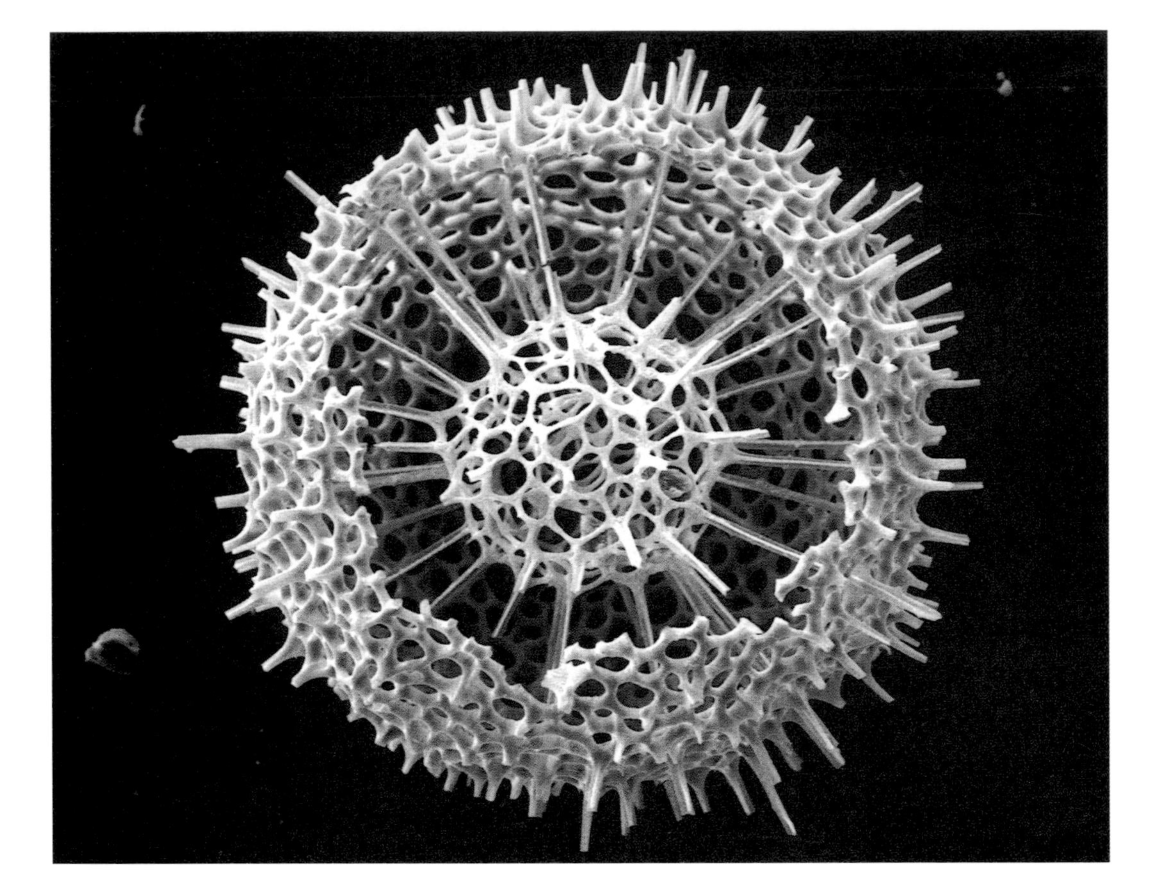

5.28 SPHERES WITHIN SPHERES: window into a radiolarian, x 300 (1, 31)

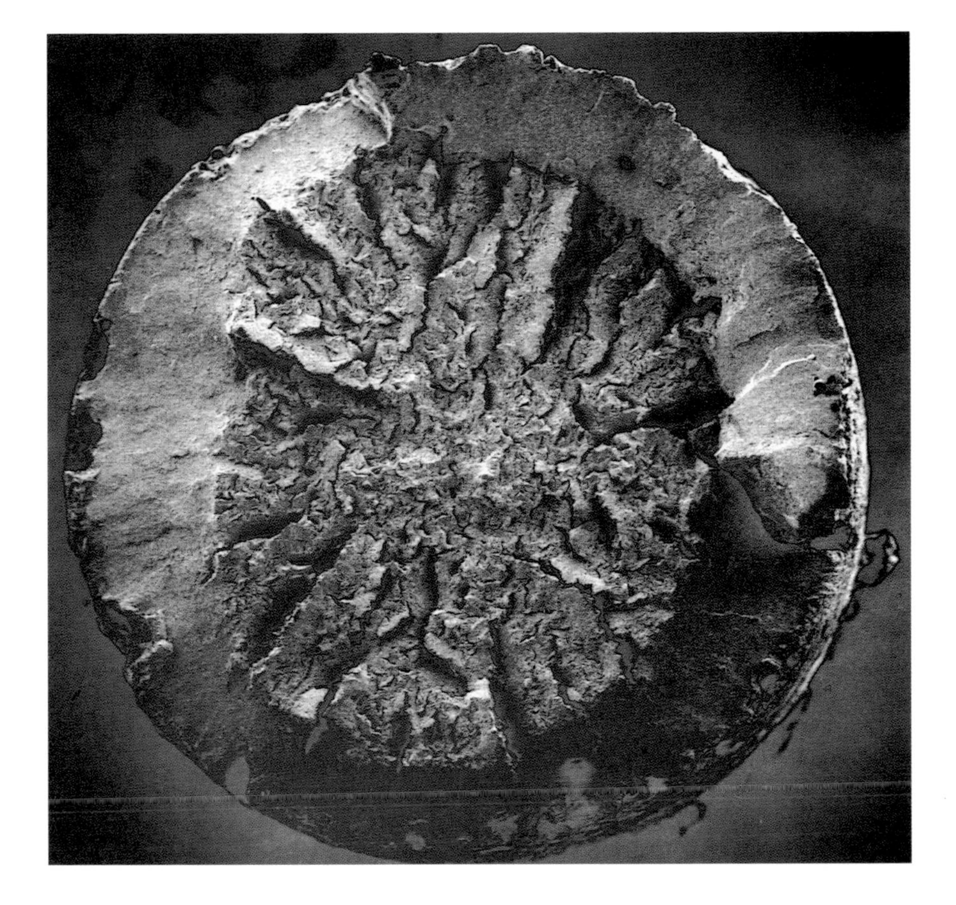

5.29 A PICTURE OF STRESS: cross-section of a broken wire from one of the cables of the Williamsburg Bridge (New York City), x 18 (57)

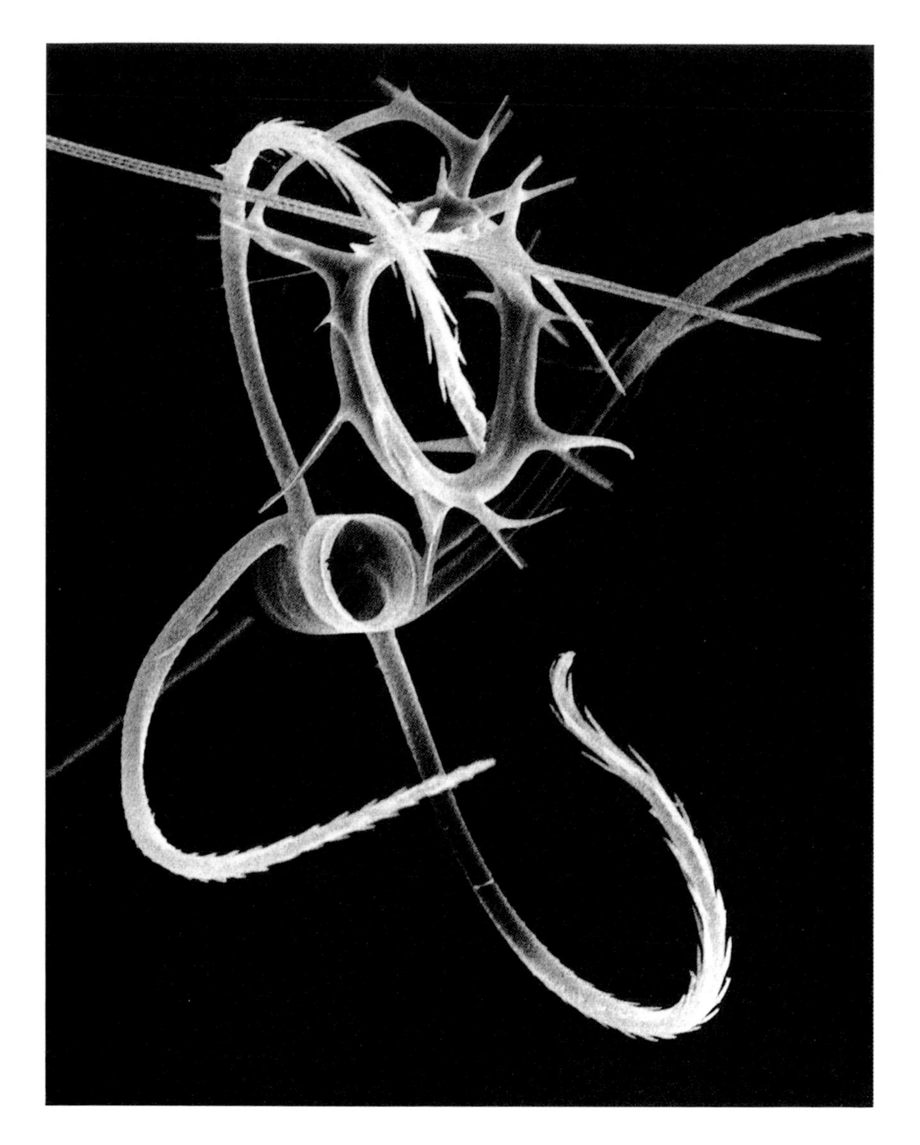

5.30 BALANCING ACT 1: a diatom and various fragments of similar sea creatures, x 500 (2, 51)

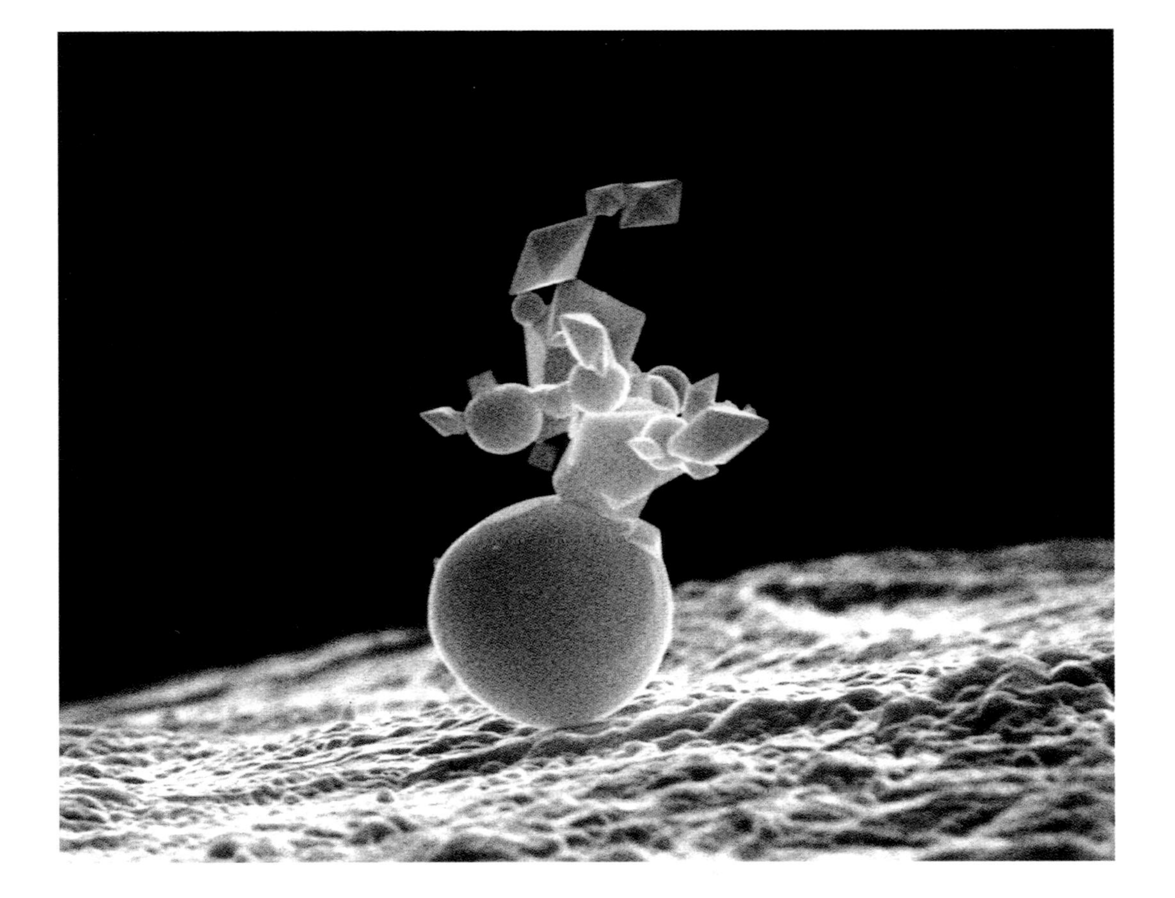

5.31 BALANCING ACT 2: at the edge of a burned tungsten wire, x 25,000 (9)

5.32 SURFACE DETAIL OF A FOSSIL FORAMINIFER: "So long, now. Thanks for stopping by!", x 5,000 (4, 58)

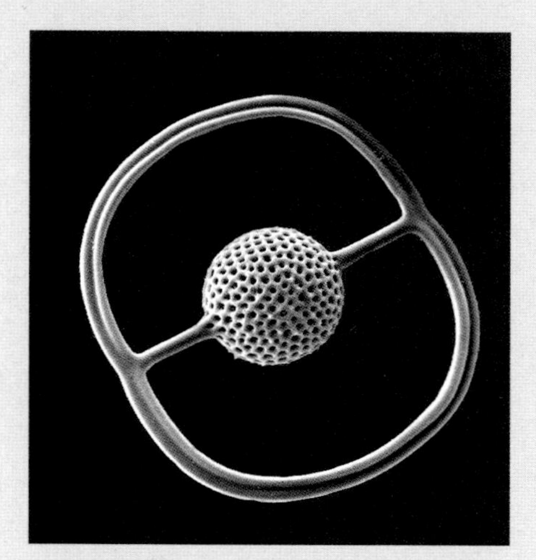

S.1 Radiolarian, x 180 (1, 51)

S.2 Diatom, x 2000 (2, 21)

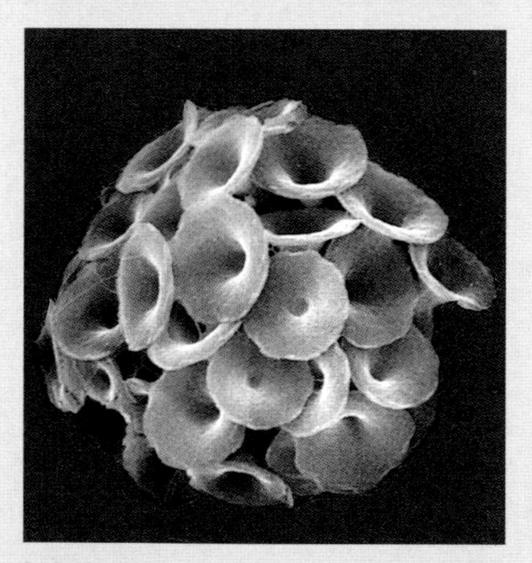

S.3 Coccolithophorid, x 3700 (3, 17)

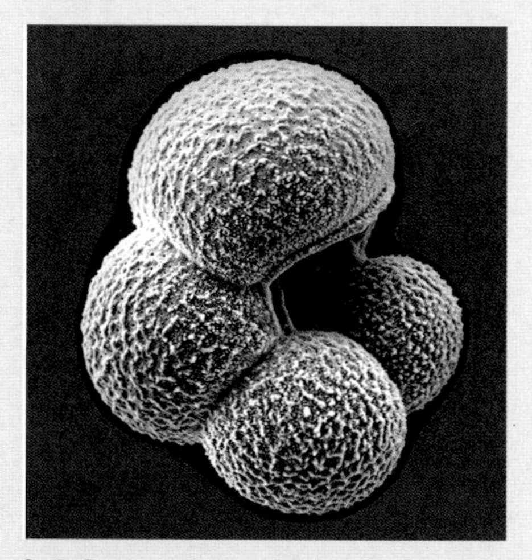

S.4 Foraminifer, x 150 (4, 62)

SOURCES & STORIES

MICROPLANKTON

▪

THESE ONE-CELLED PROTOZOA LIVE WHEREVER THERE IS MOISTURE: in the seas, in freshwater, and in soil. They consist of a heterogeneous assemblage of about 50,000 one-celled organisms whose only similarity is their membrane-bound cellular organelle. As plankton, they are either free-floaters or only weakly motile. In any case, they cannot overcome the currents that carry them. For more information, see John McNeill Sieburth, *Sea Microbes* (Oxford: Oxford University Press, 1979) and Robert D. Barnes, *Invertebrate Zoology*, chapter 2 (Fort Worth: Saunders College, 1987).

Four major groups of microplankton are represented in *Journeys in Microspace:*

1. RADIOLARIA

These floating, microscopic protozoa are distantly related to amoebae. They are found in all major oceans and reach population densities of several hundred to many thousands per cubic meter of seawater. Each radiolarian contains a single central cell body from which numerous fine sticky cytoplasmic strands radiate outward, forming a halo to capture food in the form of algae (floating microscopic plants) or other protozoa and crustacea (shrimplike floating arthropods). Some species contain algal symbionts living within the radiating strands of cytoplasm. The symbionts produce food for the radiolarian by photosynthesis.

The most remarkable aspect of these microscopic organisms is their glassy skeleton, which is created from secretions of amorphous opal. The skeleton is an ornate shell surrounding the central cell body; the opal is a form of water glass molded into unique, species-specific shapes by the radiolaria. The skeleton can be simple, consisting of scattered spicules on the surface of the cell. Other species produce very elaborate skeletons forming latticed spheres, helmet-shaped, porous shells, or ornate tripodal-shaped skeletons with richly ornamented lateral winglike extensions.

Radiolaria have inhabited the oceans for many thousands of years, and each species grows best in ocean water of a particular temperature or salinity. Upon the death of the radiolaria, their shells settle into the sediments at the bottom of the ocean and form layers, year by year. Hence, the stratified deposits of the radiolarian skeletons form a fossil record of the changing environment of the ocean over great time spans. By taking core samples of the ocean sediments it is possible to analyze the radiolarian shells in each layer of sediment and deduce what the climate of the surrounding area was for many thousands of years past. Thus, in addition to their remarkable beauty, which reminds us that even the simplest forms of life display complex patterns of behavior and construction, their fossil record is of practical importance: the fossil skeletons of radiolaria also are important during oil exploration.

O. Roger Anderson, Professor, Teachers College, Columbia University, and Senior Research Scientist, Lamont-Doherty Earth Observatory.

2. DIATOMS

These single-celled plants are found in a number of habitats, including the ocean (where they are often referred to as the grass of the sea), rivers, lakes, swamps, soils, and even on other plants and animals (they are found, for example, living on whale skins). They can be free-floating (planktonic), attached to a substrate (benthic), or found living on mud or rocks. Although a few species are capable of thriving in the dark, most of them rely on photosynthesis for life. Additionally, they draw nutrients from their environment. Because diatoms are so abundant scientists believe they play a role in controlling the carbon dioxide (CO_2) concentration of the atmosphere. Indeed, some researchers have postulated that increased productivity of photosynthetic organisms (including diatoms) in the oceans may have played a major role in reducing the levels of atmospheric CO_2 during the last glacial period 18,000 years ago.

Because diatoms are extremely sensitive to their environment, they have been used to monitor changes in physical and chemical properties of oceans and lakes. For example, diatoms sensitive to changes in lake water acidity have been used to monitor water quality and, because they are preserved in lake bed sediments, to document the recent past history of lake acidity.

Diatoms have an external skeleton (called a frustule) made of opaline silica (SiO_2+nH_2O). Because of the delicate structures produced by these skeletons, diatoms were a favored topic of study by nineteenth-century microscopists. Sedimentary accumulations of these skeletons are used to make products such as filters, toothpaste, and cosmetics. The frustule consists of two valves and has been compared to a pill box, in which one half of the box fits over the other. During the diatom's asexual reproduction, the two valves separate and secrete new valves inside the old ones. During sexual reproduction genetic information is exchanged between individuals. Marine diatom assemblages first appeared in the geological record during the waning days of the dinosaurs (the Cretaceous Period). There are individual and rare occurrences reported for the Jurassic and a possible occurrence during the Ordovician Age. In spite of these early occurrences, marine diatoms are generally regarded as having become a major marine component over the last sixty-five million years. Freshwater diatoms have a shorter history: although they probably evolved from near-shore marine forms and first appeared around forty million years ago, they became abundant about fifteen million years ago.

Lloyd H. Burckle, Adjunct Senior Research Scientist, Lamont-Doherty Earth Observatory.

3. COCCOLITHOPHYCEAE

These creatures are marine, one-celled, golden brown algae. Despite their small size, which ranges between one and fifty microns (one micron is a thousandth of a millimeter) they are one of the most important plants in the tropical through subpolar surface waters of the global ocean. Coccolithophyceae are both a major source of food for marine animals and of atmospheric oxygen. They secrete small, elaborate skeletal units called coccoliths (from *coccos*, meaning circular, and *lithos*, meaning stone) of calcium carbonate ($CaCO_3$). These units are formed within the cell and then extruded to cover the surface of the algal cell with a limestone coating. After death, coccoliths settle to the bottom of the oceans as fine-grained lime muds.

The coccolithophyceae evolved in the Jurassic Period of geologic time, about two hundred million years ago. They have undergone rapid evolutionary change, which produced elaborate variations in coccolith shape and ornamentation. This characteristic has made them important index fossils in oceanic rocks and sediments. (An index fossil is one that is well preserved, unique in appearance, and whose species had both a short life and a wide geographic range). Thus coccoliths are used to determine the geologic ages of ancient oceanic sediments and rocks. Since they first evolved, they have been a major constituent of marine sediments. Often they are the dominant form: many fine-grained limestones, like the White Cliffs of Dover, are made of coccoliths. Although their size requires electron optics to view them, the complex crystalline elaborations on geometric forms are esthetically pleasing.

Andrew McIntyre, Senior Research Scientist, Lamont-Doherty Earth Observatory, and Professor of Geology, Queens College, City University of New York.

4. FORAMINIFERA

These unicellular marine protozoans are biological relatives of the more familiar amoebae. They secrete a mineralized shell, called a test, which is generally made of calcium carbonate (lime). Foraminifera have a rich fossil record that extends back to the Cambrian Period, although they were most common and diverse in the Devonian Period. There are two different types of foraminiferal species, with either benthic (bottom dwelling) or planktonic (floating) lifestyles. Foraminifera are arguably the most valuable marine fossil group for understanding oceans from the Late Cretaceous to the modern age, i.e., the last 100 million years. They are one of the most important groups used to date the fossil record through the science of biostratigraphy. Their distributions provide information on past ocean temperatures, surface and deepwater currents, changes in water depth, the environment of deposition, and paleochemistry. Foraminiferal tests also lock away stable isotopic and trace metal information that has proven to be one of the most important means of reconstructing global changes in sea level, climate, and ocean circulation. The sequestration of their shells in deep sea sediments over the past 100 million years is an important process in the global carbon dioxide system. The interested reader should refer to "Foraminifera" by Anne Boersma, in *Introduction to Marine Micropaleontology* (New York: Elsevier Press, 1978).

Kenneth G. Miller, Professor of Geological Sciences, Rutgers University, and Adjunct Senior Scientist, Lamont-Doherty Earth Observatory.

THE MICROGRAPHS

■

5. PLATE 1.2:

A series of micrographs of a mysterious spherule that was found in a sample of freshly crushed Onondaga Limestone from New York State. The composition of this Devonian age (about 400 million years old) limestone was investigated to determine whether microscopic magnetite spherules were present in the host rock. They were, thus proving authogenesis; that is, the magnetites originated with the limestone. The goal was to investigate the mechanism that causes magnetites to reset their magnetic fields long after formation. A quick X-ray analysis during the SEM exploration of this unidentified little intruder, which was not under investigation, revealed that it is composed of calcium and zinc. Possibilities for its origin include volcanic and anthropogenic sources, but its most likely origin was cosmic. The first micrograph in this sequence is a 100 percent secondary electron image, optimized to show morphology (the object's spherical shape); the second is an electronic mix comprised of about 75 percent secondary and 25 percent backscatter images, the third is an image mix of about

25 percent secondary and 75 percent backscatter, and the last is a 100 percent backscatter image, optimized to display chemistry (showing crystals in a matrix of a different composition).

Sample and description courtesy of Susan L. Halgedahl, Associate Professor of Geology and Geophysics, University of Utah.

6. PLATES 2.1, 2.5, 2.11, 2.18, 2.21, 2.25, 2.31, 2.32, 3.3, 3.7, 3.11, 3.19, 3.23, 3.26, 3.29, 4.17, 4.26, 4.27, 5.24, 5.25, 5.26, 5.27, C.7, C.15, C.18, C.27:

The wings of a mosquito are its primary steering mechanism and the "hairs" and "feathers" are probably used for better control over its flight. In addition, mosquitoes use the sound produced by flight as a part of their courtship. The various structures on the wings may be part of a species-specific instrument that sings the appropriate courtship song. Male mosquitoes have much more plumose antennae than do females, giving them the keener sense of smell they need to locate the females to initiate courtship. Most of the hairs, or setae, on insects are sensory, usually mechanoreceptors, and can be associated with a nerve cell. In Plate 5.25, the SEM process has distorted the male mosquito: although both antennae are intact, they would look different in the living insect. The mosquitoes in these micrographs are *Aedes aegypti*, the species that carries yellow fever.

The beetle from which Plates 2.5, 3.7, and 3.26 were taken is called a "Betsy Beetle," "Bess" or "Bessey Bug," or "Patent Leather Beetle" (for its shiny black appearance). These insects live in rotting wood and make squeaking sounds. The hairs protruding from the head are probably sensory.

Scales cover the wings of most butterflies and give the wing its colors, either by pigmentation or by surface sculpturing. The ridges on butterfly wings produce metallic-looking colors by diffraction. Other colors on butterfly wings are made when blood cells dry within these structures during development. The proboscis, which the butterfly uncoils in response to food, is composed of tightly ringed sclerotic structures that allow it to coil (as in Plate 4.17) and uncoil.

The millipede in Plate 5.27, without the more pronounced mouth parts of efficient predators, lives by eating vegetables. It has excellent antennae for the senses of smell and taste, and its eyes are insignificant.

Insects and discussion courtesy of Betty Faber, Staff Scientist, Liberty Science Center, Jersey City, New Jersey.

7. PLATES 2.2, 4.16, 4.20:

A series of longitudinal views of tracheid cells in a well-preserved fossilized conifer from Tierra del Fuego, Patagonia, in each of which numerous bordered pits can be seen. The pits are the apertures that allow transmission of sap from one cell to another and thus the circulation of the tree. Plate 2.2 shows two sets of cells butting against each other. In Plate 4.16 two levels of the cells can be seen, and Plate 4.20 is from a locally deformed area of the sample.

The piece of petrified wood from which this sample was taken was given to the microscopist by Blanca and Ricardo Covacevich, sheep ranchers in central Tierra del Fuego, Chile. Description kindly provided by Gordon C. Jacoby, Senior Research Scientist, Tree-Ring Laboratory, Lamont-Doherty Earth Observatory.

8. PLATES 2.3, 4.4, 4.25:

Membrane filters and latex particles developed for industrial applications. Plate 2.3 is the corner view of a foamlike film of nylon. The film was fractured at low temperature in liquid nitrogen so that a clean break could be made. The elliptical objects are on the bottom surface of the film and appear elliptical because of the oblique view. The near-circular objects are in the cross section of the film. Plate 4.25 is from a mat of nylon 66 crystals that was prepared by spreading a thin layer of a viscous solution of nylon, formic acid, and water on a glass plate and then immersing it in a nonsolvent. The precipitation of the nylon by nucleation and growth of sheaflike crystals results in the formation of a microporous membrane. Membranes such as this one can be used to filter bacteria and other colloidal material from water. They are tough and flexible. Plate 4.4 illustrates a close-packed array of monodispersed polystyrene latex particles. A latex is a dispersion of polymer particles in water. (Latex house paint is a common example, although in this case the particles are not all the same size.) When the spheres are dispersed in water they can be used to analyze components in biological fluids and act as model colloids. The spherical particles are prepared by emulsion polymerization methods. Because the growth of the small particles begins at the same instant, all of the particles grow to exactly the same size during polymerization.

Samples and descriptions courtesy of Carl C. Gryte, Professor of Chemical Engineering and Applied Chemistry, Columbia University; latex spheres prepared by Shih-Yu Liang, Graduate Research Assistant.

9. PLATES 2.4, 2.14, 2.17, 3.4, 3.8, 3.12, 3.28, 5.31, C.24:

The microscopist attached a short piece of tungsten wire between two electrodes and sent a current through it until it flamed. Subsequent SEM exploration yielded a series of images of dendritic crystals, smooth flowing surfaces and little spheres, representing various effects of oxidation and melting. The dendritic crystals in Plates 2.4, 2.14, 3.12, and 3.28 are similar to ice crystals seen on windows during a sudden winter chill and were formed by rapid growth during condensation of the vapor phase.

Description kindly provided by Siu-Wai Chan, Associate Professor of Materials Science, Henry Krumb School of Mines, Columbia University.

10. PLATE 2.6:

Smoked whitefish served at a family reunion.

Sample courtesy of Betty Jonas, West Orange, New Jersey.

11. PLATES 2.7, 2.23:

Only two teeth like this one have been recovered from a 225-million-year-old fossil bed near Richmond, Virginia. It is believed that they came from what must have been a very tiny dinosaur, but the species has not yet been determined. Plate 2.23 is a close look at one of these teeth, revealing the intricate structure of the outer surface of the enamel. The small sharp bumps on the sides of these teeth, as shown in Plate 2.7, function like the serrations on a steak knife, enabling the animal to easily cut through the flesh of its prey. At that time in Earth's history, dinosaurs were just beginning to diversify, but the terrestrial vertebrate fauna was still dominated by synapsids, from which mammals are descendants. This fascinating fossil site, which is little more than an expanded roadside ditch, is now being threatened by a nearby housing development.

Microdinosaur tooth and description courtesy of Annika Johansson, Graduate Research Assistant, Lamont-Doherty Earth Observatory.

12. PLATES 2.8, 2.22:

Three-week-old salmon fry were investigated as a possible new animal model for microcirculatory studies, particularly intravital microscopy. Plate 2.8 is from a fin, oriented so that the top of the micrograph is toward the body of the fry where the fin is attached (not shown). The slightly raised vertical lines are arterial and venous capillaries supplying the fin with blood. Arterial capillaries bring the blood towards the tip of the fin, where it loops around and returns through the venous capillaries, terminating at bulbous junctions called capillary venules, seen here as the white spherical features. In the lower portion of the micrograph, connections can be seen between several pairs of capillaries. These are shunts that short-circuit the blood supply to non-vital areas of the body in times of crisis such as shock, rerouting blood flow to more essential areas. Shunts can be arterial-venous or venous-venous. Plate 2.22 is taken from the skin of the fry's yolk sac. Blood vessels accumulate around the small bulbous projections, which might serve to provide greater surface area for blood circulation.

Samples and descriptions courtesy of B. A. Zikria, Clinical Professor of Surgery, Columbia University, and attending surgeon at Columbia-Presbyterian and Harlem Hospital Medical Centers.

13. PLATES 2.10, 2.15, 2.28, 3.1, 3.5, 4.18, 4.39, C.3, C.16:

These microminerals are from a unique ore body in Franklin, New Jersey. In 1812, Dr. Fowler, owner of the property, recognized its unparalleled characteristics and donated collections of its minerals to Princeton and other universities. Harvard professor Charles Palache studied the mineralogy of Franklin and nearby Sterling Hill for nearly forty years in the early twentieth century. Commercial mining began in 1848, primarily for zinc but also for iron and manganese, and continued to the mid 1950s. In the course of development, the Franklin and Sterling Hill deposits were discovered to be the richest site of mineral species known in the world. Over 350 species have been found to date, some of them only in minute amounts, some known nowhere else in the world, and some comprising the most spectacular collection of fluorescent minerals known. The original deposition has been dated to about one billion years and was possibly part of an ancient rift valley. Franklin's long history, including five periods of mountain building and several periods of volcanism (which have altered preexisting minerals into new ones), has contributed to the great variety of minerals found here.

Samples courtesy of Herb Yeates, Franklin Ogdensburg Mineralogical Society; description kindly provided by John L. Baum, Curator, Franklin Mineral Museum, Franklin, New Jersey.

14. PLATE 2.12:

Buffing the flesh side of leather produces an extremely fine fibrous nap known as suede. This finish can be applied to any leather to give it suede's characteristic soft, smooth, velvety feel.

Description kindly provided by Duncan Woods, Cygnus Graphic, Phoenix, Arizona.

15. PLATE 2.13:

Chrysotile from the Setford Asbestos Mine, Setford, Quebec, Canada.

Sample courtesy of Carl Gryte, Professor of Chemical Engineering and Applied Chemistry, Columbia University.

16. PLATES 2.16, 2.26:

The fragment of rock from which these micrographs were taken is a sample of the Acasta River Gneiss outcrop in Northwest Territories, Canada. Nearly four billion years old, this coarse-grained rock was metamorphosed from part of the Earth's early crust and is the oldest known rock in the world.

Sample commercially acquired.

17. PLATES 2.19, 3.10, 4.14, 4.22, 5.2, 5.17, 5.19, C.26, S.3:

Coccolithophorids (the complete shells, composed of interleaved plates), coccoliths (the individual plates), and diatom.

Samples courtesy of Andrew McIntyre, Professor, Queens College of the City University of New York, and Senior Research Scientist, Lamont-Doherty Earth Observatory.

18. PLATES 2.24, 3.27, 4.11, C.5, C.19:

The Earth's mantle is a layer of the interior of our planet. It is the engine for many major geological processes (such as continental drift, the opening of new oceans, volcanism, etc.). The mantle lies just below the crust of the Earth and extends down for several hundred kilometers. A few years ago, some geological expeditions revealed that Zabargad Island, a small, lovely desert island in the Red Sea, is made of a block of the Earth's mantle, which originally lay very deep (over one hundred kilometers) beneath the Red Sea. This block of mantle rock, called peridotite, was lifted by compressional forces acting during the opening of the Red Sea, which is currently in the early stages of ocean formation. Plate 2.24 shows a small detail from one such rock; a crystal face from a mineral called pyroxene (a silicate of magnesium and iron) is enclosed in minerals showing tensional features. Zabargad Island has a long and distinguished history, shrouded in mystery and legend. "Zabargad" means "peridot" in Arabic, and the island has been known since antiquity as a source of gem-quality olivine crystals. Under the name of "topaz," these olivine gems are mentioned in several ancient Middle Eastern texts, including the Bible. In the book of Exodus, the High Priest's breastplate is described as containing a large peridot gem; the book of Ezekiel mentions a peridot gem. Both of these gems are probably from Zabargad. Pliny recounts how a green peridot from "the Island of Topaz" in the Red Sea was given as a gift to Queen Berenice of Egypt. This stone may well have later adorned the bosom of Cleopatra, who was one of the descendants of Berenice. Pliny describes also in his *Natural History* a sort of "rainbow mineral" with hexagonal faces, dug from a Red Sea island sixty miles from the port of Berenice, which probably corresponds to the cancrinite crystals which are also found at Zabargad. Later on, at the time of the Crusades, the Christians brought back from the Middle East many good specimens of peridot gems. These gems had mostly been robbed from Egyptian mummies buried in the catacombs of Alexandria and had probably been mined at Zabargad in Pharaonic times.

Mantle rocks well up slowly in relatively hot regions beneath the oceans, at mid-ocean ridges; when they reach a few kilometers below the seafloor, they may react with hot seawater that has penetrated into fractures and cracks of the crust. These reactions produce a family of minerals called serpentine. Plates 3.27 and 4.11 show the intricate structure of serpentine minerals in rocks dredged from the Indian Ocean floor. Plate C.5 is an image from a rock dredged in the central Atlantic Ocean by the *R/V Maurice Ewing*. It originated deep in the mantle, about one hundred kilometers below the seafloor; the vein cutting through it is made of chrysotile, a serpentine mineral composed of magnesium silicate. The small sphere shown in Plate C.19, made of pyrite, was recovered by dredge from the bottom of the Romanche Trough. The larger Romanche Fracture Zone is a huge crack running along the equator in the floor of the Atlantic Ocean from the coast of South America to that of Africa, appearing as a deep canyon, or valley, that goes almost as deep as eight kilometers below sea level, and offsets the Mid-Atlantic Ridge. The mineral aggregate in this photograph may have been formed when hot solutions loaded with chemicals (composed mainly of seawater that circulated below the seafloor and changed its composition by reacting with rocks) were discharged through the seafloor. After the discharge, the minerals cool down and some of the elements become insoluble, forming minerals such as this pyrite composed of iron and sulfur.

Samples and descriptions courtesy of Enrico Bonatti, Doherty Senior Scientist, Lamont-Doherty Earth Observatory.

19. PLATE 2.27:

Hair is made of densely compacted dead cells composed of the protein keratin. Epidermal cells, modified by a process called "cornification," form the thin, overlapping scales of keratin that serve as an external cortex surrounding a fibrous core. The hair shaft lengthens (in alternating phases of growth and rest) only from the follicle at its base, where cell division takes place. Hair can be useful as a diagnostic tool in disorders of the endocrine system.

Sample courtesy of Kathy Morrison, Croton-on-Hudson, New York.

20. PLATES 2.29, 4.8, 5.7, 5.10, 5.21:

Human cardiac and pulmonary tissue: the "corrugated" structure of striated cardiac muscle, as seen in Plate 2.29, accommodates the expansion and contraction of the beating heart. Plate 4.8 shows the interior of a pulmonary artery, where blood cells can be seen. Pulmonary arteries are elastic, to permit expansion and recoil; their walls contain numerous, circularly arranged elastic fibers, which are also apparent in this micrograph. Plates 5.7 and 5.21, from heart and lung tissue respectively, are examples of white blood cells, which (as leucocytes and macrophages) play a role in the defense of the body against infection and are prominent at sites of immune reactions. Macrophages can ingest foreign particles, microorganisms, and effete cells in the body. (Plate 5.21, of a lung infected with pneumonia, illustrates this nicely, as a motile alveolar macrophage is seen probing an airway with its pseudopod, in search of particulate matter.) Plate 5.10 shows a transverse section of a bifurcating pulmonary artery with accompanying bronchiole. The walls of bronchi and bronchioles (conducting airways) consist partly of cartilage and/or bone, which prevent their collapse during ventilation. The respiratory portion of the lung includes alveolar ducts and alveoli, which can also be seen in this micrograph. These possess an extremely thin epithelial lining that allows for the exchange of gas between air and blood.

Pulmonary and cardiac tissue samples courtesy of Marianne Wolffe, Professor of Pathology, Columbia-Presbyterian Medical Center, New York, New York and Attending Pathologist, Morristown Memorial Hospital, Morristown, New Jersey. Identification of the features kindly provided by Jerold Brett, Senior Staff Associate, and Steven Brunnert, Assistant Professor, Department of Pathology, both at the Columbia-Presbyterian Medical Center. Discussion paraphrased from Richard D. Kessel and Randy H. Kardon, *Tissues and Organs: A Text Atlas of Scanning Electron Microscopy* (San Francisco: W. H. Freeman and Company, 1979).

21. PLATES 2.33, S.2:

Diatoms courtesy of Constance Sancetta, Senior Research Scientist, Lamont-Doherty Earth Observatory.

22. PLATES 3.6, 4.13:

Iguana tail spines and skin.

Samples shed by Limbo of Jersey City, New Jersey, and provided courtesy of his owner, Mike Jones.

23. PLATES 3.9, 3.21:

Jack-in-the-Pulpit. The pollen from this plant represents either a highly evolved state or a previously unrecognized primitive condition for pollen of angiosperms (flowering plants). Similar types of pollen are produced by the Gnetales, now considered to be the closest living relatives of the flowering plants. Plate 3.9: male "flowers" on spikes, showing two to three pairs of biloculate anthers (pollen sacs) attached at the ends of short pedicels and tiny round pollen covered with short spines within the openings to each anther. Sessile (immobile) paired anthers without connective filaments are very unusual among flowering plants and may also represent a previously unrecognized primitive condition rather than the result of extreme reduction. Similar types of anthers have been found on fossils of an angiosperm-like plant that lived 216 million years ago in the southwestern United States. Plate 3.21: Five spiny inaperturate pollen grains nestled among the folds of the stigma to a female carpel.

Samples and descriptions courtesy of Bruce Cornet, former Research Associate, Lamont-Doherty Earth Observatory.

24. PLATES 3.13, C.10:

In 1948, Georges de Mestral of Switzerland returned home from a hunting trip and began to remove the thistle blossoms clinging to his clothes. Intrigued by how tenaciously they held to the cloth, he examined them under a microscope and found that they were covered with little hooks. He realized that this principle could be applied to an artificial fastening device and experimented with different materials over the next several years. In 1956 de Mestral found that if he pressed a nylon strip covered with hundreds of tiny hooks against another fashioned with hundreds of loops, they would stick together tightly but could easily be pulled apart. He named his invention "Velcro," from the French words velours (velvet) and crochet (hook). For an example of the natural model on which de Mestral's product was based, see the Enchanter's Nightshade seed in Plate 5.9.

Description kindly provided by Duncan Woods, Cygnus Graphic, Phoenix, Arizona.

25. PLATE 3.14:

Proteins are heavy, complex organic compounds composed of nitrogen, hydrogen, and oxygen, formed in the body from amino acids that themselves are formed from protein in the diet. As components of hormones, enzymes, immunoglobulins, and other substances, proteins are essential in biological processes.

Protein sample courtesy of Becker T. C. Baines, West Nyack, New York.

26. PLATE 3.15:

Tektites and microtektites are glassy fragments of molten rock formed by the cataclysmic impact of a comet or asteroid on the Earth. The microtektite shown in this micrograph is from the immense Australasian impact event that occurred about 800,000 years ago. The ejecta of this impact was dispersed over Australia, southeast Asia, and much of the Indian and western Pacific Oceans, and covered as much as one-tenth of the Earth's surface. This tiny bit of glass was recovered from a deep-sea core drilled near Indonesia and was used in a study of the possible linkage between impact events, disruptions or reversals of the Earth's magnetic field and global climate change. Evidence of a causal link was found to be insufficient to demonstrate a physical connection.

Sample and description courtesy of David A. Schneider, Postdoctoral Fellow, Woods Hole Oceanographic Institution, Woods Hole, Massachusetts, and Dennis V. Kent, Doherty Senior Scientist, and Gilberto A. Mello, Senior Research Assistant, Paleomagnetics Laboratory, Lamont-Doherty Earth Observatory.

27. PLATES 3.17, 4.6, 4.7, 4.21:

Worm tube encrustation on a shell found lodged among rocks at the base of a cliff overlooking the Sea of Cortez, Baja California. Many types of polychetes are tube-dwellers that build on hard surfaces such as shell, coral, and rock. These tubes are all made by serpulid annelid worms, also known as Christmas Tree worms, of which there are about five hundred species worldwide. They have a complicated crown, which may be a double spiral or a single loop, consisting of several feather-shaped branches. The worms make their own tubes, adding material at the wide end as they themselves grow larger. When they grow, it may look (as in these micrographs) as if this was a colony, but each tube was formed by a single worm. The worms use the crown for feeding on small particles floating in the water. Note that some of the tubes have a round cross section, while others are triangular. The tubes in these micrographs were made by several species of worms, although since the tubes of different species can be very similar to each other, only rarely can the shape be used to identify them. One will often find several very different species of these worms piling on top of each other, especially if the substrate they settled on, such as a rock or a snail shell, is in an area with high currents.

Sample courtesy of Nicky Huard, Montreal, Canada. Description kindly provided by Kristian Fauchauld, Department of Invertebrate Zoology, National Museum of Natural History, Smithsonian Institution, Washington, D.C.

28. PLATE 3.18:

Fulgurite is a mineral formed by the extreme heat of lightning boring through sand in less than one ten-thousandth of a second. This echo of a tiny lightning bolt from the sand hills of North Carolina shows that the quartz melted to form a fragile glass tube with cooler sand grains sticking along the outside.

Sample acquired commercially.

29. PLATES 3.20, C.11, C.17, C.23:

The thyroid is unusual among the endocrine glands because a precursor form of its hormone, thyroglobulin, is secreted into and stored in extracellular compartments called follicles. Thyroid hormone regulates basal metabolism, affecting every cell in the body, and is important in the development of the brain and bones of children. Plate 3.20 shows thyroglobulin actively being excreted. Production of the hormone involves subsequent reuptake of the stored thyroglobulin into the cellular lining of the follicles, which are illustrated in Plate C.11. Plate C.17 shows the end of an arteriole or venule connected to a capillary network that can be seen emerging from the thyroid tissue. Plate C.23 illustrates connective tissue from the same thyroid sample. Connective tissue is a mixture of cells, fibers, and amorphous ground substance that forms a framework of support for other tissues of the body as well as internal support for the organs. Other functions of connective tissue include transport, storage, protection, and repair.

Thyroid tissue sample courtesy of Paul LoGerfo, Professor of Surgery and Director of Surgical Oncology, Columbia-Presbyterian Medical Center, New York, New York. Identification of the features kindly provided by Steven Brunnert, Assistant Professor, Department of Pathology. Discussion paraphrased from *Tissues and Organs: A Text Atlas of Scanning Electron Microscopy* by Richard D. Kessel and Randy H. Kardon (San Francisco: W. H. Freeman and Company, 1979).

30. PLATE 3.22:

The sediment sample that included this sand grain was recovered from a core taken in the Ross Sea, off the coast of Antarctica south of Australia and New Zealand. Unlike most of the other grains in the sample, this is one of very few that exhibit a texture whose cause is as yet undetermined.

Sample courtesy of Thomas and Davida Kellogg, Professors of Geology, University of Maine.

31. PLATES 3.24, 4.10, 5.28, C.28:

Fossils of Antarctic radiolaria.

Samples courtesy of Richard Mortlock, Senior Staff Associate, Lamont-Doherty Earth Observatory.

32. PLATES 4.1, C.1, C.21:

Plates 4.1 and C.21 show the calcareous basis secreted by a modern, shallow-water sessile barnacle, *Megabalanus* cf. *tintinnabulum* (L.), that frequently fouls ships in tropical waters and gained a nearly pantropical distribution by the time of Darwin's monograph on sessile barnacles (1854). The sample was found by the microscopist on the deck of the *R/V Thomas Washington* during a 1991 leg of the World Ocean Circulation Experiment (WOCE), an international, multiyear physical and chemical oceanography expedition, near the Society Islands in the South Pacific. Several juveniles of this species were found clinging to a scutum from a large specimen of the pedunculate barnacle, *Lepas,* an oceanic genus whose species are found attached to floating objects (such as wood and pumice, but also to ships, buoys, and instruments) at or within a few meters of the sea surface. The calcareous basis illustrated in this picture is characteristic of this type of barnacle; it not only provides stronger attachment to the substratum than does the membraneous basis of more primitive forms, but the marginal mortises and tenons, formed around its margin by the radial canals, interlock with those along the basal margin of the wall, which adds considerably to the strength of the entire structure.

Plate C.1 is a detail from the exterior wall of another juvenile of this species on the same scutum. While barnacles are crustaceans and replace most of their exoskeleton at intervals by molting, unlike most other arthropods they retain a permanent outer wall or shell. The exterior of the shell is often marked by two types of growth lines, especially in juveniles. One, the "hirsute line" of little bumps and bristles illustrated here, corresponds with the molting of the cuticle of the appendages and lining of the cavity within the shell. Growth lines of the shell are formed by the secretion of calcium carbonate (calcite) and the mounds of the hirsute lines provide support for the extension of hollow whiskers (setae) that pass through from the underlying hypodermis. The setae contain living tissue sensitive to stimuli and may serve to detect currents and/or to warn the barnacle of a predator climbing its wall. This micrograph also shows the fine growth lines or ridges that generally correspond to tidal and/or related phenomena rather than the molt cycle.

An example of evolution can be seen in Darwin, who studied barnacles for eight years. In an 1846 letter to Captain Robert Fitzroy, whose company he had shared during the voyage of the *Beagle,* he wrote, " . . . for the last half-month (been) daily hard at work in dissecting a little animal (barnacle) . . . could spend another month on it, and daily see some more beautiful structure!" In a letter to colleague Joseph Hooker in 1848, he referred to barnacles as "my beloved *Cirripedia.*" By 1852, two years before his final monograph on *Cirripeds* was published, he wrote his cousin, W. D. Fox, that "I hate a Barnacle as no man ever did before, not even a sailor in a slow-sailing ship."

Identification and descriptions kindly provided by William A. Newman, Professor of Biological Oceanography, Scripps Institution of Oceanography, La Jolla, California.

33. PLATES 4.2, 4.31, C.8:

These samples were part of a research experiment investigating the feasibility of using synthetic materials as grafts in coronary bypass surgery in order to reduce complications that arise from using natural arterial grafts taken from the patient's own body. Plate 4.2 shows the endothelial lining of the arterial wall. Arterial thrombus is illustrated in Plates 4.31 and C.8, showing erythrocytes (red blood cells), platelets, and fibrin (the clotting factor). Erythrocytes sometimes stack into columns called Rouleaux Formations, one of which can be seen at the top of Plate 4.31.

Samples courtesy of John J. Castronuovo, Jr., Associate Clinical Professor of Surgery, Columbia College of Physicians and Surgeons.

34. PLATE 4.3:

Detail of a polycrystalline copper substrate from a project to develop the manufacture of multilayered Cu/Ni material. 5–500 nm-thick alternating layers of copper and nickel were investigated in the SEM to monitor the electroplating process. This method of nanoscale multilayering combines the flexibility of copper with the strength of nickel and produces a material that increases physical properties such as corrosion resistance and has ultimate applications in the electronics industries.

Sample and description by Chun-Chen Yang, Graduate Research Assistant, Department of Chemical Engineering, Columbia University.

35. PLATES 4.5, 4.35:

These micrographs are from a white spruce (*Picea glauca*) growing along the Coppermine River in Northwest Territories, Canada. Evidence from several trees in the region indicates that the summer of 1641 became extremely cold in the middle of the growing season, causing all growth processes to stop. Plate 4.35 illustrates the effects on the growth ring for that year in one of these trees. The wood cells formed, but were weak due to the lack of lignification (the process that cements cell walls much as mortar strengthens brick walls). The cells later broke and were displaced as the tree swayed in the wind, as can be seen within the narrow growth ring (third from left) for that year. The following year, 1642, also had either an abbreviated or abnormally cold growing season, as indicated by its narrow ring (second from left). The rings on either side of these two are of normal width and structure, suggesting normal growing seasons for those years. This tree was stronger than others in the region, which have circular breaks at about 1641.

Plate 4.5 shows radial intertracheid pits in single rows, some showing the torus (a valve that can close off the pit or move aside to leave it open); in others the torus was removed as the surface was split away.

Samples and descriptions courtesy of Gordon C. Jacoby, Senior Research Scientist, Tree-Ring Laboratory, Lamont-Doherty Earth Observatory.

36. PLATE 4.9:

Huon pine is a relic tree species that flourished widely during the Miocene Era, ten to twenty million years ago. Today it is only found in the temperate rain forest of western Tasmania. Huon pine lives longer and grows more slowly than almost any other tree species in the world. Its documented lifespan exceeds 2,500 years of age and its average annual tree-ring width rarely exceeds 0.5 mm per year. The wood is virtually impervious to decay and scientists have found well-preserved logs as much as 38,000 years old. The wood in this image is from some 2,000 years ago. The annual tree rings from living Huon pine trees and subfossil wood are now being used to produce a continuous record of past environmental changes for Tasmania that will eventually extend back to the last glacial period.

Huon pine sample and description courtesy of Edward R. Cook, Research Scientist, Tree-Ring Laboratory, Lamont-Doherty Earth Observatory.

37. PLATE 4.23:

Backscatter image of a flat polished sample of intimately mixed celadonite (the light portions of the image) and calcite (the dark portions of the image), suggesting that the two minerals were cogenerated, at least in part. This example of tholeiite is a product of seawater-rock interaction under certain conditions (including relatively low temperature) and represents less than 10 percent of the flows. The rock was erupted above sea level when the land mass that became Norway and Greenland began to split apart during Magnetic Anomaly 24, estimated at between fifty-three and fifty-six million years ago. Anomaly 24 is the oldest seafloor-spreading magnetic anomaly identified in the Norwegian-Greenland sea.

Sample courtesy of Anne P. LeHuray, Principal Geochemist, Ebasco Environmental. Description paraphrased from "Rb-Sr Systematics of Site 642 Volcanic Rocks and Alteration Minerals," A. P. LeHuray and E. S. Johnson, *Proceedings of the Ocean Drilling Project, Scientific Results*, Vol. 104, 1987.

38. PLATES 4.24, 5.12, 5.14:

Calcium oxalate urolithiasis is a pervasive medical problem, resulting in significant pain, lost productivity, and morbidity. Despite many years of intensive research into this problem, a reliable therapeutic inhibitor is still lacking. Extracorporeal shock wave lithotripsy (a nonsurgical treatment developed in the early 1980s, in which focused external shock waves break up a urinary tract stone *in situ*) may not be a completely effective treatment for kidney stones. At a recent National Institute of Health Consensus Development Conference on Urolithiasis, it was concluded that the search for medical prevention for stones must continue. Thus a major objective of this research was to study the mechanism of crystallization of calcium oxalate, looking at how solution variables determine the calcium and oxalate crystallization products. One question asked during this experiment was, "What are the implications of these results for clinical efforts to reduce calcium and oxalate concentration in the urine?" A model was also created to study the mechanism of polymeric influence on calcium oxalate growth rate, which was used to extract data, such as the growth rate constants, from the measured growth kinetics. Plate 4.24 illustrates the crystals formed during one experiment, showing a mixture of calcium oxalate monohydrate and calcium oxalate dihydrate. Plate 5.12 is a single crystal of pure calcium oxalate monohydrate grown under a different set of variables. Plate 5.14 is a backscatter image from a natural kidney stone, showing the process of its formation in the body: calcium phosphate and calcium oxylate crystals have settled out of solution in concentric layers that are seen in this micrograph as alternating dark and light bands.

Samples and descriptions courtesy of Joseph Manne, Department of Urology, Long Island Jewish Medical Center.

39. PLATES 4.28, 4.29:

Submarine canyons dissect continental margins of the world's ocean basins. The severe erosion of the continental slopes leads to the formation of these submarine valleys, some of which attract the attention of marine geologists. Because of the many environments in which canyons evolve, numerous hypotheses have been postulated to explain their origins. Their development is related to subaerial erosion during episodes of emergence, to a direct connection with a river mouth or delta, to scouring of canyon floors and walls by submarine currents or turbidity currents, to biologically induced failures, and to dissolution of carbonate rocks by corrosive fluids.

Expeditions with submersibles were conducted to explore submarine canyons in the New Jersey continental margin in order to understand how these canyons formed. In the New Jersey lower slope the canyons are incised into siliceous biogenic chalks, which are rocks formed from the remains of microfossils. The tests of the microfossils are of opaline silica (opal A) and calcium carbonate. With progressive burial and increased temperatures, the biogenic opal A of the siliceous tests transforms into opal CT. The transformation involves the expulsion of substantial amounts of water (20 percent water content). Expulsion of the fluids leads to a volume reduction and to fracturing of the bedrock.

To a great extent submarine canyon development in the New Jersey lower slope is controlled by the silica diagenetic transformation of the sediments. Canyon excavation is initiated by turbidity currents and mass-wasting. Once canyon erosion is started, it continues through positive feedback. Excavation releases stress, and the fractures in the extensively fractured sediment expand. Erosion proceeds with the spalling and sliding of blocks. Large chalk blocks have been found on the continental rise, far away from their source.

Plate 4.28 shows detail from a diatom in the canyon chalk that is still well preserved, as illustrated by the smooth edges on its areoli (holes). It can be contrasted with the detail in Plate 4.29 from a diatom test made of opal A that is starting to undergo its diagenetic transformation, as shown by the etched rims of the areoli.

Samples and description courtesy of Cecilia McHugh, Assistant Professor, Queens College of the City University of New York, and Visiting Research Scientist, Lamont-Doherty Earth Observatory.

40. PLATE 4.30:

Eggshell is formed of calcite, one of the crystalline forms of calcium carbonate. During development of the egg, two shell membranes, one inner and one outer, are laid down around the albumen. A sparse matrix of protein runs through the crystals of the shell, which is attached to the outer membrane by hemispherical structures known as mammillary knobs. Anchoring fibers run from the outer membrane into the knobs, which have protein cores. The knobs are probably calcified soon after formation, before the shell enters the shell gland, and subsequently act as nuclei for the growth of the calcite crystals comprising the shell. The calcification of the knobs is thought to involve the binding of calcium ions to the organic cores of the knobs by means of the sulfonic acid groups on the acid-mucopolysaccaride-protein material of which the cores are composed. The main part of the shell is called the palisade layer and is composed of columns of tightly packed calcite crystals; the columns extend from the mammillary knobs to the cuticle. Occasional pores, through which the embryo takes in oxygen and gives out carbon dioxide during incubation, run up between the crystals in spaces formed where groups of knobs come together.

Discussion paraphrased from T. G. Taylor, "How an Eggshell is Made," *Scientific American*, March 1970.

41. PLATES 4.32, C.20:

Seawater samples were filtered on board the *R/V Endeavor* during a 1989 biological oceanography expedition in the North Atlantic Ocean. This expedition was one segment of an international multiyear expedition known as the Joint Global Ocean Flux Study (JGOFS), investigating the flux of carbon in the world's oceans. Plate 4.32 shows an intriguing deposition of sea salt onto the filter, along with the expected organisms. As the filter (substrate with holes) dried, crystals of magnesium carbonate or calcium carbonate precipitated on a scale too fine to be seen or chemically detected by the X-ray analyzer attached to the SEM. However, they attracted ions of sodium and chlorine—salt—

which then precipitated in the characteristic shape of the attracting, but invisible, carbonate crystals. The circular patterns are "drying horizons" formed as the salt, which was both visible and chemically detectable, came out of solution. Plate C.20 is an example of what was more expected. It is a coccolithophorid, collected from surface water south of Iceland.

Filtrate samples courtesy of Ray Sambrotto, Research Scientist, Marine Biology Department, Lamont-Doherty Earth Observatory.

42. PLATE 4.33:

Enlargement of the cuticle, or skin, of a 220-million-year-old leaf from Little Conewago Creek in York County, Pennsylvania. This leaf came from a Late Triassic conifer that lived within the great Newark Supergroup rift valley just before Africa and North America began to break apart and form the Atlantic Ocean. The conifer grew along the shores of a giant lake that filled the Gettysburg basin and probably represents some of the food eaten by the first herbivorous dinosaurs. The tiny spines on the surface of its leaves helped the plant conserve moisture during the extended dry season of a tropical monsoon climate by reducing air flow velocities and creating air turbulence at the leaf surface.

Sample and description courtesy of Bruce Cornet, former Research Associate, Lamont-Doherty Earth Observatory.

43. PLATE 4.34:

Internal dentin within the root of a deciduous (primary) lower left second molar. The dentin is being lost through a process called resorption, by cells known as odontoclasts, to make way for the emerging succedaneous (permanent) tooth. The large craters are probably the sites where the odontoclasts sat as they were eating away at the dentin. The small holes are the openings of the dentin tubules, which are channels through the dentin that are connected to the internal pulp chamber of the tooth.

Sample courtesy of Miyaca Two Dogs (age 11). Interpretation kindly provided by Marc A. Rosenblum, Associate Clinical Professor, University of Medicine and Dentistry of New Jersey.

44. PLATES 4.36, C.24:

Backscattered electron images of two examples of very rapid crystallization of silicate magma under laboratory conditions. Olivine (the dark portions of the image), with the composition Mg_2SiO_4, forms dendrites due to more favorable growth from silicate liquid (the light portions of the image) along the edges and corners, rather than faces, of crystals at high rates of cooling. (The small bright spots dispersed throughout these images are crystals of molybdenum metal that precipitated during quenching of the sample from very high temperature.) Similar textures, referred to as spinifex, are commonly encountered in ultramafic lavas known as komatiites. This magma type is found in ancient (Archaean) geologic terrains (more than 2.5 billion years old), but is virtually absent in the more recent geologic record. Geologists believe these magmas reflect the earlier period of Earth's history when the interior temperatures were substantially higher, and a globally encircling ocean of magma may have existed at depth.

Samples and description courtesy of Charles E. Lesher, Associate Professor of Geology, University of California at Davis.

45. PLATE 4.37:

Found in arthropods, particularly in insects and crustaceans, compound eyes are arrangements of numerous long cylindrical lenses (called ommatidia) packed together. Each lens can receive light and form a separate image. The compound eyes of diurnal insects are specialized to form a mosaic image, while those of nocturnal insects form an overlapping series of images. Though an insect's vision is less sharp than that of other creatures, its compound eyes give it the advantages of detecting movement over a large field of view and the ability to respond quickly to sudden movements.

46. PLATE 4.38:

Intrahepatic stones are uncommon, though when they do occur (generally in Asian adults) they are associated with deficiencies in childhood diet.

Sample courtesy of Tsung-Hung Peng, Oak Ridge, Tennessee.

47. PLATE 5.1, C.13:

When a pollen grain lands on the stigma of a flower, enzymes in the stigma cause formation of a pollen tube that grows from the pollen grain, traverses a relatively long path through the style of the flower, and finally penetrates the tissue of the ovule. The tube's function is to facilitate fertilization by delivering the male gametophytes carried within the pollen grain to the female gametophytes, born within the ovary of the flower. In this micrograph from a goldenrod sample collected outside the microscopist's home, the pollen tube has just begun to grow along the surface of the stigma.

Description kindly provided by Sarah J. Fowell, Research Associate, University of South Carolina.

48. PLATES 5.3, 5.9:

Plate 5.3 is a seed from a *Stellaria pendula* (*Caryophyllaceae* family) plant, collected in Holland; Plate 5.9 is the seed of a *Circaea lutetiana* (*Onagraceae* family), collected in 1947 in Watoga State Park, West Virginia. Both seeds are components of an extensive international reference seed collection given by Uhl Kuhn (then age 94) to Dr. D. M. Peteet while she was a graduate student at New York University. The collection, which duplicates Kuhn's life collection for the Department of Agriculture in Newark, New Jersey, has proven essential to the palynological/macrofossil studies at Lamont and the Goddard Institute for Space Studies (GISS).

Samples and description courtesy of Dorothy M. Peteet, Research Scientist, GISS, New York, New York, and Adjunct Research Scientist, Lamont-Doherty Earth Observatory.

49. PLATE 5.4:

Testate amoeba, as the name implies, are microscopic amoebae that produce a test, or shell, surrounding them. Testate amoebae occur mainly in freshwater environments such as lakes and streams. Others dwell in soil to several inches depth, with different species occurring at each depth. The shell surrounding the testate amoeba is composed of an organic layer secreted by the surface of the amoeba, and forming at first a sticky cement wherein small fragments of mineral matter are embedded before the organic cement hardens. The shape of the shell and the kind of mineral matter deposited in it, when present, are characteristic of the species. Some species produce flattened almond-shaped tests with overlapping scales of silica. Others have a more globose test containing closely spaced grains of mineral matter. The opening of the test is sometimes elaborate with lobes, toothlike extensions around the periphery, or other ornamentations. Testate amoebae are important members of the freshwater and soil microbiotic community, and their presence may be essential to a healthy and productive agricultural soil environment.

Sample and description courtesy of O. Roger Anderson, Professor, Teachers College, Columbia University, and Senior Research Scientist, Lamont-Doherty Earth Observatory.

50. PLATE 5.5:

Stones in the pancreas are formed primarily from calcium carbonate (as calcite), proteins, and polysaccharides. The shape of the stone depends on its location and extension to the side branches of the ducts. Pancreatic calculi are white or dull brown in color, usually with a knobby surface. The protein plug at the core is composed of a very fine network of fibers. Pancreatic concretions are associated with diabetes, chronic alcoholism, tropical pancreatitis (probably due to malnutrition), and other causes, and have been a subject of investigation since the seventeenth century. In 1920, M. Barron described the association of pancreatic stone disease and diabetes, suggesting the possibility of a hormone being secreted directly into the lymph, which had control over carbohydrate metabolism. This observation is widely believed to have guided F. G. Banting and C. H. Best to the discovery of insulin in 1921.

Sample and description courtesy of C. S. Pitchumoni, Professor of Medicine/Community and Preventive Medicine, New York Medical College, and Chief of the Division of Gastroenterology, Our Lady of Mercy Medical Center, Bronx, New York.

51. PLATES 5.6, 5.8, 5.13, 5.30, C.9, C.12, S.1:

These radiolaria and diatoms were collected near Barbados during a study of radiolarian ecology to document the climatic conditions where modern radiolarian species occur. This helps scientists relate the distribution of living species to that of fossil species deposited during geological time spans.

Samples courtesy of O. Roger Anderson, Professor, Teachers College, Columbia University, and Senior Research Scientist, Lamont-Doherty Earth Observatory, and Dr. Atsushi Matsuoka, Professor of Earth Science, Nigata University, Nigata, Japan. Description courtesy of O. Roger Anderson.

52. PLATE 5.11:

Gypsum was found in shallow water carbonates (tufa) from a fossil shoreline of Pyramid Lake, Nevada, about 100 meters above the present day lake level and was probably formed during evaporation of lake water some time after the lake retreated during the last glacial cycle, about 13,000 years ago. This sample was part of a search for the distribution of detritus material in an attempt to correlate the distribution of thorium. This project uses uranium-thorium and carbon-14 dating methods to study the late Pleistocene hydrological balance of two closed basin systems in the Great Basin of the western United States, the Lahontan Basin (of which Pyramid Lake is the largest residual lake) and the Owens River system. From the study of fossil shorelines, the lake level at Lake Lahontan in the late Pleistocene fluctuated up to more than 150 meters above today's lake surface. The wet periods in the western Great Basin have been found to be likely correlated with the cold periods in the North Atlantic Ocean and the El Niño-related abnormal climate response. The Great Basin is now located in a rain shadow blocked by the Sierra Nevada mountains, where the average precipitation today is usually below 200 millimeters per year.

Sample and description courtesy of Jo C. Lin, Graduate Research Assistant, Lamont-Doherty Earth Observatory.

53. PLATE 5.15:

Volcanic glass shards from an eruption that occurred on the North American continent approximately ten million years ago. This ashfall is responsible for the death of a unique assemblage of Miocene rhinos, camels, horses, and birds who were gathered at a watering hole (in what is now Nebraska) at the time of the eruption. The animals died from inhalation of these glass shards and were quickly buried in ash. Their skeletons are incredibly well preserved; some even contain stomach contents and unborn young. The ash responsible for this death assemblage is believed to have erupted in Idaho, more than 700 miles to the west. The fossil site, Ashfall Fossil Beds State Park, is still being excavated by a team headed by Dr. M. Voorhies of the University of Nebraska and is open to the public.

Volcanic ash sample and description courtesy of Janet Saltzman, Graduate Research Assistant, Lamont-Doherty Earth Observatory.

54. PLATES 5.16, 5.20:

Microscopic marine plants such as these diatoms take up atmospheric carbon, which is a major factor in global warming. Plant growth in the ocean is proportional to the availability of dissolved iron in dust transported from the continents by winds. Because there is little source for dust in ice-covered Antarctica, there is less marine plant life in the Southern Ocean than in the other oceans; thus there is less uptake of CO_2. An experiment was developed and carried out in 1990 by the late John Holland Martin in which he fertilized fresh seawater samples from the Ross Sea (near the Antarctic continent south of Australia and New Zealand) with iron. Martin's controversial hypothesis was that the added iron would increase the abundance of marine plant mass and result in increased uptake of CO_2, which, if performed on a large scale *in situ*, could result in atmospheric reduction of CO_2. These diatoms are from a suite of samples recovered during the Ross Sea experiment. For further discussion of this experiment, see *Discover Magazine*, April 1991.

Samples courtesy of John H. Martin, former Director of Moss Landing Marine Laboratory, Moss Landing, California.

55. PLATE 5.22:

These joined proteins, which may be thyroglobulin, were found in a fine-needle aspiration biopsy of a mixed papillary/follicular carcinoma from a human thyroid. (The patient survived and went on to produce *Journeys in Microspace*.)

Sample courtesy of Michael P. Scherl, Otolaryngologist, Westwood, New Jersey. Identification by Steven Brunnert, Assistant Professor of Pathology, Columbia-Presbyterian Medical Center.

56. PLATES 5.23, C.4, C.6:

These species of planktonic foraminifera are from the uppermost layers of deep-sea cores, representing the Holocene Epoch (the last 10,000 years). These same species can be found living today in the overlying water column. *Globorotalia menardii* (Plate C.4), is a tropical planktonic taxon that was absent in the Atlantic during the last glacial maximum, although it was present there during the previous (and present) interglacial periods, and *Sphaeroidinella dehiscens* (Plate C.6) is a marker for the last four million years.

Samples courtesy of Tsunemasa Saito, Professor of Geology, Yamagata University, Yamagata, Japan. Descriptions by Kenneth G. Miller, Professor of Geological Sciences, Rutgers University, and Adjunct Senior Scientist, Lamont-Doherty Earth Observatory.

57. PLATE 5.29:

Fracture surface of a wire specimen taken from Cable D of the Williamsburg Bridge, connecting Brooklyn to Manhattan. The wires in this bridge were not galvanized, an exception to the general practice, and many of the outer wires in each cable showed rust. There are four main cables with 7696 wires per cable, each wire of nominal diameter 0.0192 inches. In 1987–88, at the request of a joint Commission of the City and State of New York and under the supervision of the consulting engineering firm of Steinman, Boynton, Gronquist, and Birdsall, the cables were opened at five locations. At each location, samples were taken at several depths and eight angles around the cable. The samples, 20 to 40 feet in length, were taken to the Carlton Laboratory of the Department of Civil Engineering at Columbia University, where smaller test specimens were chosen in a statistical fashion. The appearance of the fracture was one of the criteria used to evaluate the condition of the wires. This specimen was pulled to failure in a laboratory testing machine. The fracture started near the center of the wire in the manner of a ductile fracture with a very fine roughness, and became coarser as it spread out, presumably rather rapidly—the noise of breaking is both loud and sharp—until the spreading changed to a shear mode on a slant near the surface.

Sample and description courtesy of Daniel Beshers, Professor of Metallurgy, Henry Krumb School of Mines, Columbia University.

58. PLATE 5.32:

Surface detail of a fossil Antarctic foraminifer, showing pores and carbonate pustules. This sample was one of several checked in the SEM to monitor the effectiveness of a cleaning procedure. The cleaned forams were used in a trace metal study as part of a project investigating paleochemistry and paleoclimate through the ice ages.

Sample courtesy of Flip Froelich, Senior Research Scientist, Lamont-Doherty Earth Observatory.

59. PLATE C.2:

This 201-million-year-old pollen grain, *Corollina meyeriana*, was recovered from rocks that lie on the boundary between the Triassic and Jurassic periods of Earth history. *Corollina* pollen was produced by a variety of conifers that are now extinct but were particularly abundant in the Jurassic period. This specimen was recovered from rocks that were formed in a series of large lakes that existed along the eastern coast of North America during the Late Triassic and Early Jurassic. Collectively, these rocks are known as the Newark Supergroup. This particular species of pollen allows the location of the Triassic/Jurassic boundary to be identified with some precision in the Newark Supergroup, as it increases in abundance at the boundary. The Early Jurassic explosion of *Corollina* pollen is due in part to the regional extinction of numerous Late Triassic pollen species (and, by extrapolation, the plants that produced them). Following the disappearance of the Triassic species, spore-producing ferns dominated the landscape for a brief interval, followed by the rise of the *Corollina* producers. This pattern of extinction and subsequent temporary expansion of ferns is characteristic of geologically rapid mass-kill episodes. Hypothetical causes of these catastrophic extinctions include dramatic changes in climate and impact of a large comet or meteorite.

Sample, description, and computer color-enhancement courtesy of Sarah J. Fowell, Research Associate, University of South Carolina.

60. PLATE C.14:

In the mid-sixteenth century, breeches evolved into separate items of clothing for the legs that were knitted like gloves. Their manufacture was revolutionized with the invention of the stocking loom in the late sixteenth century. Shortly afterward silk stockings began to be produced in France, and cotton stockings appeared by the mid-seventeenth century. The invention of nylon in 1938 by DuPont chemist Wallace Carothers radically transformed the hosiery industry and market; the first nylon stockings were placed on sale in New York City on May 15, 1940. Within a few hours over four million pairs were sold.

Description kindly provided by Duncan Woods, Cygnus Graphic, Phoenix, Arizona.

61. PLATE C.25:

Small jaws still bearing teeth, such as this one, were recovered in Virginia during a study of Triassic-Jurassic fossils from the Newark Supergroup of eastern North America. Roughly 225 million years ago this microscopic jaw, about 5 mm long, sat in the mouth of a small reptile as it chewed on plants and insects along the shores of a lake. Future work will determine the species. Many new animal fossils have been recovered from this remarkable site, including teeth from the earliest known venom-injecting reptile (*Uachitodon kroehleri*) and the teeth of a tiny freshwater shark (*Lissodus paulus*).

Microjaw sample courtesy of Annika Johansson, Graduate Research Assistant, Lamont-Doherty Earth Observatory, and Hans-Dieter Suess, Adjunct Associate Research Scientist, Lamont-Doherty Earth Observatory, and Associate Curator, Department of Vertebrate Paleontology, Royal Ontario Museum, Ontario, Canada. Description by Annika Johansson.

62. PLATE S.4:

Polynyas are little-understood transient features of open water within what is otherwise solid pack ice in the Earth's polar regions. Spanning time scales on the order of years or decades, polynyas are warmed by unknown causes and are areas of high biologic productivity. The multiyear expedition, NEW WATER Polynya Project, is investigating an Arctic polynya. The living planktonic foraminifer in this micrograph, *Neogloboquadrina pachyderma,* was collected in 1992 with a plankton tow by the *R/V Polar Sea* from about 150–200 meters depth in the water column off the northeast coast of Greenland.

Sample and description courtesy of Karen Kohfeld, Graduate Research Assistant, Lamont-Doherty Earth Observatory.

63. PLATES C.3, C.8, C.20, C.25, C.28:

Computer color-enhancement by Andrew Paul Leonard.

Columbia University Press
New York Chichester, West Sussex

Library of Congress Cataloging-in-Publication Data

Breger, Dee.
Journeys in microspace: the art of the scanning electron microscope / Dee Breger
p. cm.
Includes bibliographical references (p.).
ISBN 0–231–08252–5
1. Photomicrography. 2. Scanning electron microscopy.
3. Photography, Artistic. I. Title
QH251.B73
778.3' 1—dc20 94–37829
CIP

Casebound editions of Columbia University Press books are printed on permanent and durable acid-free paper.

Printed in Hong Kong

c 10 9 8 7 6 5 4 3 2 1

Designers: Leah Lococo and Karin Martin
Text: Centaur and Geometric 415
Compositors: Leah Lococo and Karin Martin
Printer: Oceanic Graphic Printing
Binder: Oceanic Graphic Printing